Hippolyte Fizeau, physicist of the light

James Lequeux
Emeritus Astronomer at the Paris Observatory

17, avenue du Hoggar
Parc d'Activité de Courtabœuf, BP 112
91944 Les Ulis Cedex A, France

Printed in France

© **2020, EDP Sciences**, 17 avenue du Hoggar, BP 112, Parc d'activités de Courtaboeuf, 91944 Les Ulis Cedex A

This work is subject to copyright. All rights are reserved, whether the whole or part of the material is concerned, specifically the rights of translation, reprinting, re-use of illustrations, recitation, broad-casting, reproduction on microfilms or in other ways, and storage in data bank. Duplication of this publication or parts thereof is only permitted under the provisions of the French Copyright law of March 11, 1957. Violations fall under the prosecution act of the French Copyright law.

ISBN (print): 978-2-7598-2045-0 - ISBN (ebook): 978-2-7598-2188-4

Introduction

The first half of the nineteenth century saw the emergence in France of physicists and astronomers of extraordinary quality, who established classical physics to a large extent and even created astrophysics. The preceding century had seen great precursors like Lagrange, Laplace, Lavoisier or Monge, while prestigious institutions had promoted science and education; but, thanks the social promotion made possible by the Revolution, by the creation of the Polytechnic school by the Convention, and by the Egyptian campaign, new talents of the highest order emerged. The revolutionary and romantic enthusiasm allowed them all boldness. The most important physicists and astronomers of this generation were undoublty Ampere, Arago, Carnot, Fourier, Fresnel and Malus. Scientific biographies have been published on each of them, at least in French. A second, no less brilliant generation includes Fizeau, Foucault and Le Verrier. The last two have been, too, the subjects of recent scientific biographies in English. Only Fizeau remained little studied, which is why I wrote the present book. Later, in the second half of the nineteenth century, French science was on the decline, and it would be difficult to quote physicists and astronomers with as much genius as the preceding ones, with the exceptions of Poincaré, Langevin and the Curie. The reasons for this decline, at a time when science flourished in Germany, England and the United States, remain to be studied in detail.

Fizeau's life was that of a wealthy bourgeois, safe from material concerns, who was able to concentrate fully and without hindrance on science. He was generally described as a "landlord" or "annuitant" in official documents. The private part of his life offers no particular interest, and is certainly not the focus of this book. As for his scientific career, until 1849 it merged with that of Leon Foucault, with whom he worked almost constantly. They had much in common: the same age, medical studies, considerable manual skill and peerless inventiveness. In 1849, Fizeau carried alone his famous measurement of the velocity of light; then, the following year, the two men fell out with each other after their competition to measure the difference in the velocities of light in air and water. They now followed divergent paths, often harsh and unexpected in the case of Foucault who died young, in 1868, of what was likely multiple sclerosis. For his part, Fizeau made other brillant experiments, then became an established and considered scientist, well inserted in his time, which did not prevent him from having occasionally some bright and innovative ideas. He died in 1896 after a long career, and a particularly long service as a member of the Academy of Sciences, which he had entered on January 2, 1860.

I want to thank my wife Geneviève and my friend William Tobin for their careful reading of the French text and their numerous suggestions for correcting and improving it. William Tobin also communicated or showed to me unpublished documents. I thank Florence Greffe for her warm welcome to the Archives of the Academy of Sciences and for allowing me to reproduce and publish numerous autograph documents from the Fizeau funds. Similarly, the Mayor of Suresnes and Mrs. Marie-Pierre Deguillaume, director of the Museum of Urban and Social History of this city, have authorized me to publish other autograph documents as well as photographs of instruments that belonged to Fizeau: many thanks to them and to the museum staff. Marie-Christine Thooris kindly provided photographs of the replica of the Fizeau apparatus that belongs to the museum of the Ecole Polytechnique. Finally, I thank my colleagues of the Library of the Paris Observatory, who are always ready to make available to users images from their collections.

Contents

Introduction	III
Chapter 1. The beginning of a scientific life	**1**
1.1 The daguerreotype	3
1.2 A decisive encounter	8
Chapter 2. A fruitful collaboration	**15**
2.1 Interferometry and the nature of light	16
2.2 The nature of the infrared radiations	22
Chapter 3. The Doppler-Fizeau effect	**27**
3.1 Christian Doppler	28
3.2 Christophorus Buijs-Ballot	33
3.3 Hippolyte Fizeau	35
3.4 The future of the Doppler-Fizeau effect	38
Chapter 4. The velocity of light and electricity	**45**
4.1 A fertile period	46
4.2 The first direct measurement of the velocity of light	48
4.3 The measurements of the velocity of light after Fizeau	55
4.4 The velocity of electricity	58
Chapter 5. The "crucial experiment": the velocity of light in air and water	**65**
5.1 The project of Arago	66
5.2 Fizeau and Foucault take up Arago's experiment	68
5.3 The steeple chase	71
Chapter 6. The drag of æther	**75**
6.1 Act 1: Michell, Arago and Fresnel	76
6.2 Act 2: Fizeau and Michelson	81
6.3 Act 3: Fizeau and Michelson again	89
6.4 Act 4: Lorentz, Einstein and von Laue	94
Chapter 7. The diameter of stars	**97**
7.1 A brilliant idea	98
7.2 The first tests	103
7.3 Michelson again!	108
7.4 Stagnation and renewal of astronomical interferometry	112

Chapter 8. A highly esteemed scientist **115**
    8.1   A shortened family life 116
    8.2   Some secondary but innovative works 118
    8.3   A pillar of French physics 122
    8.4   A studious end of life 125

Appendix 1. Genealogy of Fizeau and his wife **127**

Appendix 2. Chronology **129**

Appendix 3. Correspondence Fizeau-Foucault **131**

Bibliography **135**

Index **139**

Chapter 1
The beginning of a scientific life

On 1839 August 19, Arago presents to the Academy of Sciences and the Academy of Fine Arts the photographic process of Daguerre. Bibliothèque de l'Observatoire de Paris.

Armand-Hippolyte-Louis Fizeau, sixth in a family of nine children, several of whom died in infancy, was born in Paris, 7 rue Thibautodé in the Louvre district, on the 23rd of September 1819. He was baptized on September 29th in the Saint-Germain l'Auxerrois church. His sister Gabrielle, a religious person, died in 1862. The two other sisters and the two younger brothers remaining died in 1854, 1880, 1880 and 1885 respectively (see Appendix 1). Little is known about his mother, Béatrix (or Béatrice) Marie Petel. His father, Louis-Aimé Fizeau (1776-1864), had married her on 20th June 1809. A few years after the birth of Hippolyte, in 1823, his father was appointed a professor at the Faculty of Medicine of Paris, during the reorganization of the faculty which resulted in the expulsion of several of its members for political reasons (some of the excluded people are famous in France: Pelletan, Pinel, Vauquelin). He was himself dismissed for similar reasons in 1830, during the July Revolution[1]. Linked to Laennec (1781-1826), he held the chair of internal diseases; he was one of the first adept of auscultation of patients.

Hippolyte studied as a non-resident at Collège Stanislas and was supposed to succeed his father as a physician. But his health, affected by severe headaches, forced him to interrupt his medical studies. Soon, he recovered sufficiently to be able to follow his scientific tastes. An autograph sheet[2], unfortunately difficult to read, indicates the course of his studies and the name of his teachers (the boxes are in the original):

1835: Rhetoric

1836: Also philosophy

1837: Chemistry (Dumas, Boussingault, Orfila), Physics (Dulong)

1838: Travel; [illegible]

1839: [illegible] Sickness; Geology courses, physical studies, dissections, hospitals, chemistry, Mr Magendie; Good work ; Ideas on a Memoir on the shape of drops.

1840: Trip to Le Havre; Gold salts; Lectures by Mr Élie de Beaumont; [illegible]; Course of Dr. Blainville. Daguerreotype .

1841: Travel in Anjou; Bromine; electroplating; Lectures by Mr. Blainville, Regnault, Élie de Beaumont.

1842: Travel; Note on bromine; Course of Regnault; Mathematics.

---

[1] Corlieu, A. (1896) *Centenaire de la Faculté de Médecine de Paris*, Paris, Imprimerie nationale.

[2] Académie des Sciences/Institut de France, fonds 64 J, Hippolyte Fizeau, dossier 9.28.

The teachers[3], who are almost all still famous today, were the best of the time. Alfred Cornu (1841-1902), a disciple of Fizeau that we will encounter later, said[4] that Fizeau also attended the course of Popular Astronomy of François Arago (1786-1853). Although we have found no other trace of this, it should not surprise us because all the intelligentsia of the time pressed for these courses, whose success was enormous. They were given at that time in the College de France, then from February 1841 in the amphitheater of 800 seats that Arago had built at the Observatory. Always generous with young people, Arago remarked Fizeau and closely followed his first works, which he often mentioned in an eulogistic way at the Academy of Sciences. However, the course that had most intrigued Fizeau was that of Regnault, who taught optics at the *Collège de France*. There is no doubt that it confirmed Fizeau's choice of research career.

1.1 The daguerreotype

We have just seen that in 1839, while he was continuing his studies, our young man began to think about research. A first Memoir on the shape of drops of water came to nothing, but he now embarked in a completely new field: photography. On the 7th of January 1839, Arago, who was the *Secrétaire perpétuel* (permanent secretary) of the Academy of Sciences, presented to the Academy the process of Louis Mandé Daguerre (1787-1851), an invention soon to be known as the daguerreotype[5]. Working with Nicéphore Niepce (1765-1833) until the latter's death, Daguerre submitted a copper plate covered with a thin, polished layer of silver, to iodine vapor, forming silver iodide on the surface of the plate. Upon exposure to light, the silver iodide was more or less reduced to metallic silver in exposed areas. The plate was developed in mercury vapor at 60-80 °C. The excess silver iodide was dissolved in a sodium thiosulfate solution, and the plate was washed and dried. The positive image was formed by silver-mercury amalgam grains that scattered light in the most exposed areas,

---

[3] Jean-Baptiste Dumas (1800-1884) and Jean-Baptiste Boussingault (1802-1887), famous chemists; Mathieu-Joseph-Bonaventure Orfila (1787-1853), chemist, toxicologist and medical examiner; Pierre-Louis Dulong (1785-1838), physicist; François Magendie (1783-1855), physiologist; Léonce Élie de Beaumont (1798-1874), chemist and geologist; Henri-Marie Ducrotay de Blainville (1777-1850), zoologist and anatomist; Victor Regnault (1810-1878), chemist and physicist.
[4] Cornu (1897), p. C2. See this reference in the bibliography.
[5] *Arago, F. (1839), *Comptes rendus hebdomadaires de l'Académie des sciences (CRAS)* 8, p. 4-7.

giving the whites and grays, while the specular reflection on the bare polished silver gave the blacks.

During his presentation, Arago, who had just listed the immense possibilities of photography, proposed that it fell into the public domain. His request was heard: on the 9th of July 1839, the Chamber of Deputies voted an annual pension of 8,000 francs to Daguerre and 4,000 francs to the heirs of Niepce, and on August 19th the details of the process were disclosed by Arago during a joint session of the Academy of Sciences and the Academy of Fine Arts. Immediately, scientists, engineers and enthusiasts worldwide started making daguerreotypes. Daguerre took the opportunity to sell the necessary equipment, while the optician Charles Chevalier (1808-1895) manufactured cameras.

However, the results were somewhat disappointing: while the images were beautiful, they were also very fragile, and the exposure times were so long that there was almost no question of photographing living characters. Then our young Fizeau, who was just twenty, intervened. He brought decisive improvements to the daguerreotype. To fix the image and make it more brilliant, he prepared a solution of sodium and gold thio-sulfide, by mixing solutions of gold chloride and of sodium thio-sulfide. He coated the plate with this product and heated it: gold replaced silver in the amalgam of the grains, and the bare silver that formed the base of the plate was covered with a thin layer of gold that browned it, making deeper blacks. Most daguerreotypes that remain today were treated by this method, that Fizeau published in 1840 in the *Comptes Rendus* (Proceedings) of the Academy of Sciences[6].

Somewhat later, he thought to expose the plate to vapors from a very dilute solution of bromine for a few moments, before exposure[7]. The silver iodide was then replaced by silver bromide, more sensitive to light: the exposure time was reduced to twenty seconds in bright light, allowing portraits of living people to be taken. Fizeau also remarked that one could reduce the exposure time further by increasing the aperture ratio of the photographic objectives, which was only f/15 in those sold by Daguerre or Chevalier; indeed, the objectives

---

[6] *Fizeau, H. (1840) *CRAS* 11, p. 237-238. Soon after, Arago presented a daguerreotype obtained in this way to the Academy, "which differentiates from all other similar trials by its remarkable perfection and also by the no less remarkable fact that the daguerrian image was not at all altered."

[7] *Fizeau, H. (1841) *CRAS* 12, p. 1189-1190.

with several lenses built by various opticians soon reached a larger aperture ratio.

Fizeau worked with Noël Paymal Lerebours (1807-1873), the optician of the Observatory of Paris, who was also a great photographer: he made many daguerreotypes, and he was the author of a treatise of photography that had at least five editions. The fifth edition[8] has as a co-author his collaborator Marc Secretan (1804-1867), who also became a famous optician and succeeded him. They inserted notes by Fizeau (Fig. 1.1). With Fizeau's bromination process and very open objectives, Lerebours said that he could obtain poses of less than one second.

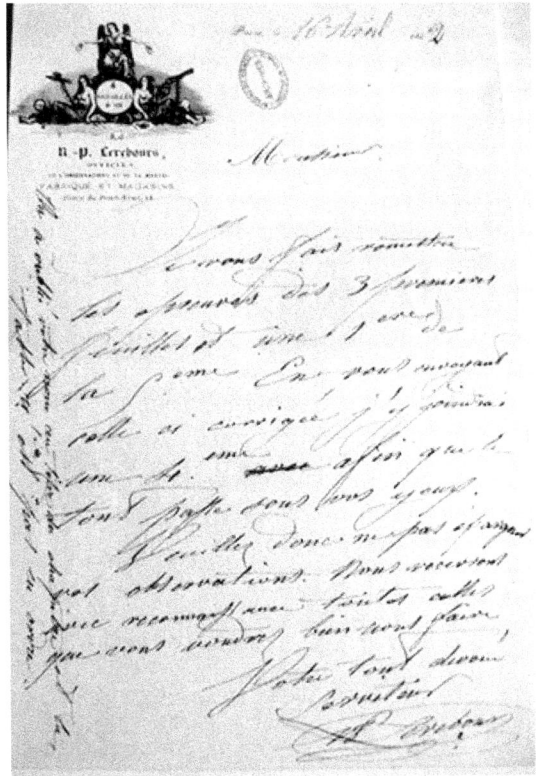

FIGURE 1.1 – Letter from Lerebours to Fizeau about inserting an article by the latter in his treatise of photography, on the use of bromine to sensitize daguerreotypes. Académie des sciences-Institut de France, Fonds 64 J, Hippolyte Fizeau, dossier 8.02.

---

[8] Lerebours & Secretan (1846).

As there was a lot of money to earn with daguerreotypes, competition was fierce, especially between Lerebours and Chevalier, who also had published a treatise of photography in 1841[9]. They did not fail to inveigh against each other by interposed writings, in particular about a new development that looked promising: the reproduction of photographic images. Indeed, the daguerreotype was a unique item, and it would have been interesting to replicate it. Fizeau developed an electroplating method that implied electro-deposition of a copper layer on the daguerreotype[10]. His contemporaries said that the layer came off pretty easily. Then Fizeau obtained another layer in the same way from the negative plate, which was a copy of the original[11].

Fizeau was not the only one to have this idea and to put it into practice: a man named Krasner claimed the priority as early as the 2nd of November 1840[12], and Chevalier said in his treatise (Chevalier, 1841, p 60-66) that he had obtained conclusive results in January of 1841; the weekly journal *L'Artiste* also mentioned his success on February 7th[13]. As for Fizeau, he made his presentation to the Academy of Sciences in March 1841.

In 1843-1844, Fizeau used another method to replicate the daguerreotypes, a work undertaken for a competition run by the Society for the Encouragement of National Industry, with a prize of 2,000 francs[14]: he attacked with an acid the daguerreotype plate, which was then sacrificed until it was sufficiently excavated. A fairly complex treatment allowed him to obtain "a considerable number of copies"[15]. Three plates in an important book entitled *Excursions daguerriennes* were obtained by this process from daguerreotypes by Lerebours[16].

---

[9] Chevalier (1841).
[10] *Fizeau, H. (1841) *CRAS* 12, p. 401-402.
[11] *Fizeau, H. (1841) *CRAS* 12, p. 509.
[12] *(1840) *CRAS* 11, p. 712.
[13] *L'Artiste* (1841) 2d Series, t. 7, 6, p. 94. One reads: "The galvano-plastic instrument was made with much care by two optical engineers, MM. Chevalier and Lerebours, who rival in talent and skillfulness [...]. M. Charles Chevalier obtained a result in his first trials [...], the application of the metal was so exact that a daguerreotype plate was reproduced in spite of its light and shallow lines."
[14] Fizeau did not obtain the prize, and 1,000 francs went to MM. Choiselat and Ratel: *Bulletin de la Société d'Encouragement pour l'Industrie Nationale*, 52e année (1853) p. 297, accessible via http://cnum.cnam.fr.
[15] *(1843) *CRAS* 16, p. 408; *Fizeau, H. (1844) *CRAS* 19, p. 119-121.
[16] *Anonymous (1840-1843) *Excursions daguerriennes : vues et monuments les plus remarquables du globe*, t. 2, Paris, Rittner & Goupil.

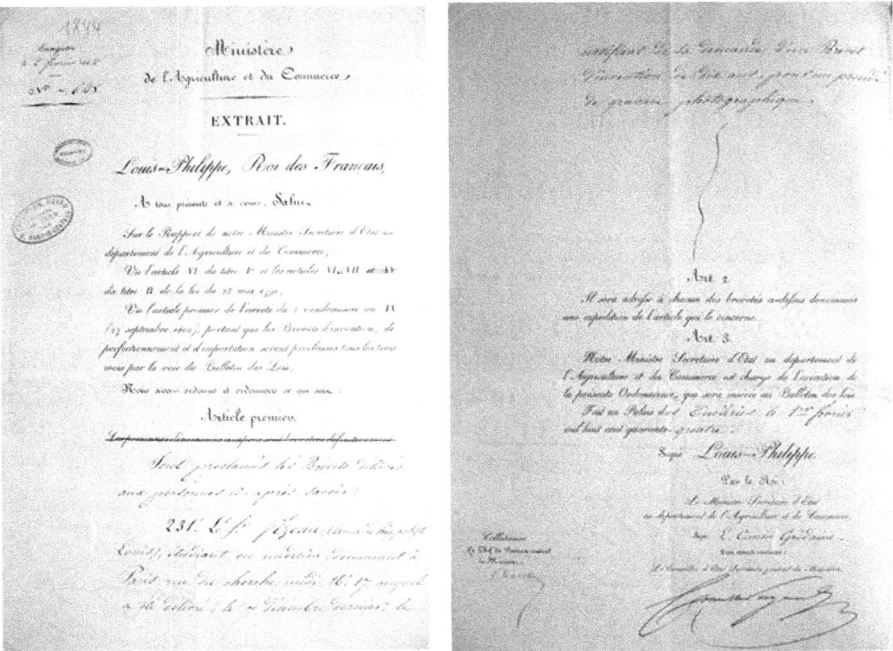

FIGURE 1.2 – A patent issued to Fizeau for reproduction of daguerreotypes by electroplating (front and back). Académie des sciences-Institut de France, Fonds 64 J, Hippolyte Fizeau, dossier 8.02.

Facing the competition, Fizeau filed a patent that was recorded on the 2nd of February 1844 (Fig. 1.2). It is therefore likely that some priority was acknowledged to him, because he was the only one to have published in the *Comptes rendus*; Arago, who held him in high esteem, was perhaps no stranger to this. Taking a patent proved an unnecessary precaution, because the process was complicated and risky, especially since "most often, the original piece is in relief; so it is necessary to produce a counter-proof and to submit it to the same operations [electroplating] to produce a relief plate" (Chevalier 1841, p. 66). It was soon abandoned: after William Henry Fox Talbot (1800-1877), "its use was abandoned because of the large uncertainties that it included, who used the patience of the experimenters."[17] The future was in the neg-

---

[17] Article by Fox Talbot in Atheneum, dated the 4th of April 1853, accessible via http://foxtalbot.dmu.ac.uk/letters/transcriptFreetext.php?keystring=Fizeau&keystring2=&keystring3=&year1=1800&year2=1877&pageNumber=13&pageTotal=15&referringPage=0

ative-positive process by contact onto paper invented by Fox Talbot, that he patented in 1841 under the name of *calotype*. Fizeau himself congratulated Fox Talbot for this progress[18]. The daguerreotype then had only ten years to live[19].

1.2 A decisive encounter

The young Fizeau was now famous, and many photography enthusiasts came to ask for his advice. This was the case of a young man of the same age, Léon Foucault (Box 1.1 and Fig. 1.3), who was doing experiments on the daguerreotype on the side. The two young men probably had known each other at the Stanislas College or at the School of Medicine. Everything brought them together: a social environment of good bourgeoisie, parallel studies of medicine, and especially the love of science.

FIGURE 1.3 – Léon Foucault around 1862. Bibliothèque de l'Observatoire de Paris.

---

[18] Letter from Fizeau to Fox Talbot dated the 7th of August 1844, accessible via http://foxtalbot.dmu.ac.uk/letters/transcriptFreetext.php?keystring=Fizeau&keystring2=&keystring3=&year1=1800&year2=1877&pageNumber=5&pageTotal=15&referringPage=0

[19] For an history of photographic reproduction, see Daniel (1995).

Box 1.1. Léon Foucault (1819-1868)

Léon Foucault was sort of a double of Fizeau: they were born five days apart, they both began medical studies but turned to physics, and both had a huge imagination and great experimental skill. The difference was that Foucault was basically a mechanic while Fizeau had more of a taste for and skills in mathematics. Both started their first scientific experiments at their own expense, but Foucault's resources were more limited so he had to make a living by writing scientific articles for the *Journal des débats* from 1842, indeed with great erudition and critical sense. Made famous by the work done in common with Fizeau or by himself, and especially by the famous pendulum experiment, which dates from 1851, Foucault became a protégé of Napoléon III who imposed his appointment as "Physicist of the Observatory" to Le Verrier in 1855. He quarreled with Fizeau and continued his first-class scientific work alone: the design and construction of the first modern telescopes with a silvered glass mirror from 1857, then in 1862 the first accurate measurement of the velocity of light, with a rotating mirror. Rapidly weakened by a disease that was probably multiple sclerosis, he died in 1868.

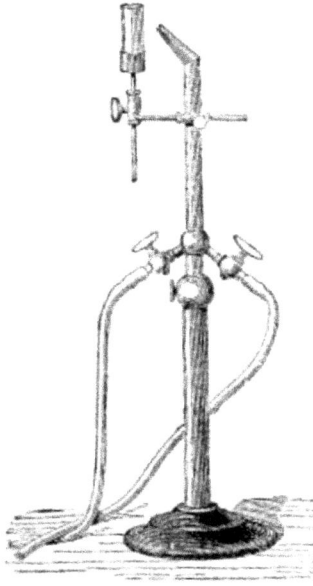

FIGURE 1.4 – The Drummond's light (limelight). The light is produced by a piece of chalk heated by a torch which burns a mixture of hydrogen and oxygen. This light source was still used at the end of the nineteenth century: the image is taken from a catalog of scientific instruments dating back to 1891. Conservatoire Numérique des Arts et Métiers.

Foucault probably contacted Fizeau for details on his method of bromination of plates. He perfected it and described his process in a booklet of ten pages, immediately published in the practical guide of Charles Chevalier[20]. This note ends with a compliment to Fizeau:

> "But I must say in closing, that the important idea, the capital idea belongs to Mr. Fizeau,: to renew the solution [of bromine in water] for each plate."

The two men worked separately until the second half of 1843, when they decided to collaborate. Their first collaboration was about the application of the daguerreotype to photometry, that is to say, the measurement of the intensity of light. This project had already been proposed by Arago in 1839, when he described what could be expected of photography before the Academy of Sciences[21]:

> "The physicists succeed quite well in determining the comparative intensities of two lights adjacent from each other and perceived simultaneously; but there are only imperfect means to make this comparison, when the simultaneity condition does not exist. [...] I don't hesitate to say, that the reagents [sic] discovered by Mr. Daguerre will hasten the progress of one of the sciences that honor most the human spirit. With their help, the physicist will now proceed by way of absolute intensities; he will compare the lights by their effects. If useful, the same table will give him the prints of the dazzling rays of sunlight, of the rays three hundred thousand times fainter of the moon, and of the light of the stars. These impressions, he will equalize them either by weakening the stronger lights with excellent means, the result of recent discoveries[22], or by letting the brightest rays act only for a second, for example, and the others for up to half an hour when needed."

Fizeau and Foucault attempted to compare the intensity of sunlight with that of two other light sources. The purpose was utilitarian: what was the best source of light to be used for laboratory experiments? Was it the sun, or the electric arc invented by Humphry Davy (1778-1829) in 1801, or the oxy-hydrogen light of Thomas Drummond (1797-1840), actually invented by Goldsworthy Gurney (1793-1875)[23] (Fig. 1.4)? To

---

[20] Chevalier (1841).
[21] *Arago, F. (1839) Le Daguerréotype, *La France littéraire* 35, p. 404-420.
[22] This obviously is for Arago the attenuation achieved by turning relative to each other two polarizers crossed successively by the light.
[23] For Drummond's light, see Lauginie, P. (2013) *Drummond's Light, Limelight: a Device in its Time,* accessible via http://www.scientificinstrumentsociety.org/bulletin-127/

do this, they produced the image of the three sources on daguerreotypes plates successively with an achromatic lens. In order that these images were just at the threshold of sensitivity of the plate after development under mercury vapor, they played with different parameters: the distance from the source to the lens, the focal length of the lens, a diaphragm limiting the useful diameter of this lens, and even the exposure time.

The result[24] was that the brightness of the incandescent chalk of Drummond's light was fainter than that of the positive carbon electrode of the arc, which was itself weaker than the Sun. If the Sun, taken as the reference in good conditions, had a brightness of 1,000, that of the arc went from 136 to 385 according to the difference in potential applied to the electrodes, and that of Drummond's light only 0.5 to 6.85 according to the gas pressure in the torch. This referred to the luminance in the blue-ultraviolet light to which the daguerreotype is sensitive: this is what Fizeau and Foucault called the *chemical intensity*, because these are the wavelengths that produce the chemical reaction of the reduction of silver salts, which is the basis of photography. They realized that the relationship between the brightnesses could be different for the light to which the eye is sensitive, and they decided to measure visually the ratio of the brightnesses of the Drummond's chalk and the arc, comparing the images formed on a screen by differently diaphragmed lenses. They found that this ratio was not much different from that obtained with the daguerreotype. They concluded mistakenly that "the measures of chemical intensity [...] relative to sunlight, to the electrodes of the arc, and to a mixture of oxygen and hydrogen projected on lime, would also be the measures of the optical intensities of these light sources." This was approximately true when comparing the electric arc to the Drummond light, but not for the Sun whose temperature is much higher. At the end of the communication of Fizeau and Foucault, Arago reminded of his "experiments, already very old, in which he compared, by direct photometric means, the light of the Sun with that of the arc."[25]

This measurement does not seem very remarkable to us, but it was a first, and it required significant financial means to produce the oxy-hydrogen light and especially the electric arc: no less than 138 Bunsen batteries were purchased for this purpose.

---

[24] *Fizeau, H. & Foucault, L. (1844) *CRAS* 18, p. 746-755 and p. 860-862.
[25] We have not found in the Memoirs on photometry of Arago any reference to these experiments, which could have been performed in 1815, a year when Arago made various photometric experiments.

Was it then that Arago told Fizeau and Foucault about a problem that preoccupied him: the darkening of the edge of the solar disk? Indeed, he wrote that he had obtained a daguerreotype of the Sun on which he found "that the rays that come from the central part of the disc of the Sun have a stronger photogenic action than those from the edges."[26] This referred perhaps to the first successful astronomical photograph, that of the partial solar eclipse of the 15th of March 1839, which had long been present at the Paris Observatory but is lost today. The question was of importance, because the edge darkening necessarily implied that the Sun's light came from a glowing gas and not from a liquid or a solid: the brightness of an incandescent solid or liquid depends very little on the angle of incidence, only the polarization of the emitted light, while that of a gas whose temperature varies with depth, as is the case for the solar atmosphere, depends on the incidence while there it is no polarization. Arago had already concluded that the Sun was an incandescent gas in 1811, noting the absence of polarization of the light from the edges of the solar disk. Yet the very existence of the darkening of the edges was controversial, the measurements apparently giving conflicting results. Fizeau and Foucault worked to obtain daguerreotypes of the Sun. For this, the solar light was reflected horizontally by a heliostat to a lens, at the focus of which the daguerreotype was placed. But the Sun is so bright that the exposure time was to be between 1/60 and 1/100 of a second: it was not possible to use the usual method of a cover removed and replaced manually on the lens. Fizeau and Foucault imagined an "original enough" shutter consisting of a plate with a horizontal slit of appropriate width, which they dropped in front of the camera: this was the ancestor of the curtain shutter.

Many daguerreotypes were obtained in this way. All show a darkening of the solar edge. The only large one that is preserved is in the reserves of the French *Musée des Arts et Métiers/CNAM*. The image, which shows several sunspots, is 91.5 mm in diameter and was obtained at the focus of an achromatic lens with a focal length of 9.88 m. This is probably the first beautiful photograph of the Sun. A reproduction was later engraved in Arago's *Astronomie populaire* (Fig. 1.5).

---

[26] *Arago, F. (1854-1862) Œuvres complètes, t. 10, p. 231-250.

The beginning of a scientific life 13

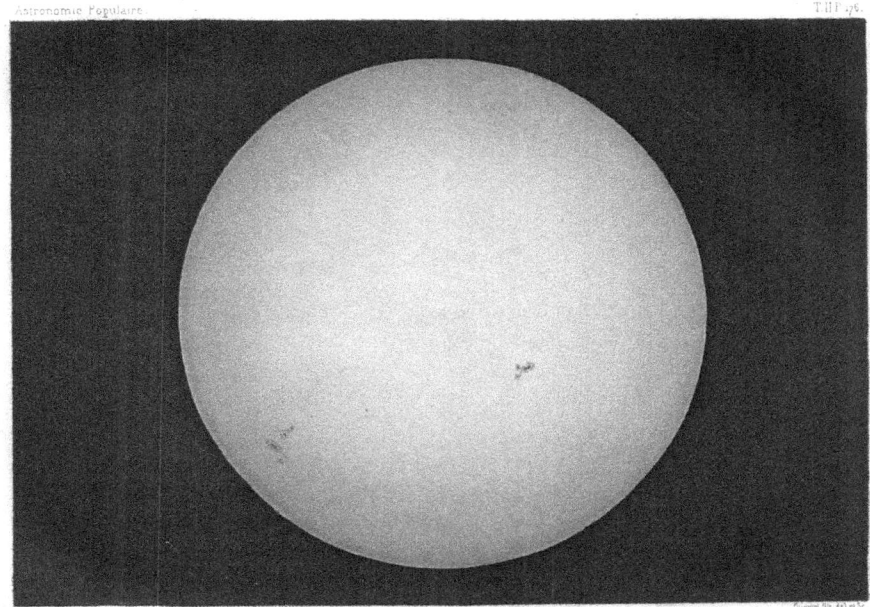

FIGURE 1.5 – Daguerreotype of the Sun obtained by Fizeau and Foucault on the 2nd of April 1845 at 9:45 am, reproduced in Arago's *Astronomie populaire*, t. 2. The sunspots are outlined by a line, an addition of the engraver. The limb darkening is faithfully reproduced. Author's collection.

Fizeau and Foucault did not submit their results on the limb darkening of the Sun to the Academy. Their contemporaries, in particular Father Angelo Secchi (1818-1878)[27], concluded from this darkening that the luminous layer of the Sun, called the *photosphere* by Arago, was surrounded by an absorbing layer, which absorbed more solar radiation at the edges because a thicker path was then crossed by the light. This idea survived for a century, until realistic models of the Sun's atmosphere were developed by the German Karl Schwarzschild (1873-1916) and his successors since 1906.

Foucault et Fizeau continued for some time to make daguerreotypes[28], and to work together on different projects.

---

[27] *Secchi, A. (1875 et 1878) *Le Soleil*, 2 t., Paris, Gauthier-Villars: voir t. 1, p. 199.
[28] One can see a few reproductions of Fizeau's daguerreotypes and photoengravings on http://gallica.bnf.fr

Chapter 2
A fruitful collaboration[29]

A Silbermann's heliostat that belonged to Fizeau, and that he probably used for his optical experiments with Foucault. The plane mirror was driven by a clockwork mechanism located in the cylindrical box, so that the light received from the Sun was reflected horizontally in a fixed direction by compensating for the diurnal motion. The instrument was set each day as a function of the height reached by the Sun at midday. Heliostats were widely used in the nineteenth century because they were the most powerful light sources for optical experiments and photographic enlargements. Musée d'histoire urbaine et sociale de Suresnes, inv. 997.00.1568.

---

[29] This chapter owes much to William Tobin, who has studied this collaboration in depth in his book (Tobin 2003).

2.1 Interferometry and the nature of light

In 1845, the controversy on the nature of light that had been raging since the beginning of the century was beginning to subside. Newton had proposed that light was made of particles with mass, and had explained refraction by an acceleration of these particles when light entered a refracting medium. Following their experiments on diffraction and interference, Thomas Young (1773-1829) and Augustin Fresnel (1788-1827) had seriously undermined this theory and proposed to replace it by the wave theory, which dated back to Christiaan Huygens (1620-1695). However, Newton's reputation was so great that it was difficult to admit that he could have been wrong, so the French physicists were still divided between supporters of the corpuscular hypothesis, as Jean-Baptiste Biot (1774-1862), and supporters of the wave hypothesis, the leaders of which were André-Marie Ampère (1775-1836) and Arago.

It is in this context that Fizeau and Foucault, both of whom were very familiar with interference experiments (Young's holes or slits, or Fresnel's mirrors), raised the following question: up to what difference in path length between two light rays originating from the same source, as in these experiments, can we observe the interference phenomenon?

In the Young's slits experiment, two narrow parallel slits cut in an opaque sheet are illuminated by a point source of light. Interference fringes between the two beams from these slits form on a screen. In white light, a bright central fringe is observed, corresponding to equal optical paths (zero path difference between the two rays), and on each side one or two fringes are seen with less contrast and some color. Then everything blurs, and away from the central fringe we only see a uniform illumination. When narrowing the spectral range of the light, more fringes are seen. Certainly this experiment had "put beyond doubt the existence of [a] periodical constitution [...] at the origin of wave motion; but the blurring of fringes when the path difference of the interfering beams exceeded a few tens of wavelengths, suggested that the regularity vanished after a few dozen oscillations; the rapid extinction of this regular regime of light sources put into doubt the valuable analogy they offered with sources of sound, and even became worrisome for a correct explanation of certain phenomena of diffraction."[30]

---

[30] *Cornu (1897) p. C.6.

A fruitful collaboration

Such was the motivation of our two young scientists; but to answer the question, they had to work with monochromatic light: they would then have expected to see many equidistant fringes on either side of the central fringe, until they eventually disappeared at large path differences.

The problem at the time was that there was no monochromatic light source that would have allowed them to observe these unscrambled fringes up to large path differences. At least, such was the opinion of our two young scholars, but they had not yet realized that sodium vapor produces emission lines favorable to their experiment, and did not know that in Scotland David Brewster (1781-1868) had built a lamp emitting these lines with high intensity in 1823. They got around this difficulty in an extremely ingenious way. It was to select, in an interferometry experiment using white light, the light that falls on the screen at a large distance from the central fringe, and to analyze it with a prism. This principle is illustrated in Figure 2.1.

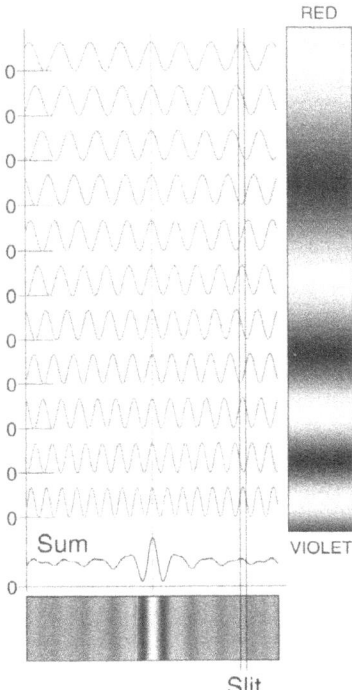

FIGURE 2.1 – Principle of the experiment of Fizeau and Foucault. The light source emits in a broad range of wavelengths. An interference device such as Young's slits or Fresnel's mirrors produces interference fringes at each wavelength of this light whose spacing is proportional to the wavelength (some are shown schematically by sinusoids). Their superimposition produces a central fringe and some lateral fringes: this is what we see on a screen in an interference experiment (bottom). By selecting a small portion of the jumbled pattern with a slit cut in the screen, as shown, and by analyzing the transmitted light with a spectroscope, a modulated spectrum with alternate bright and dark stripes is obtained by interference (right). Note how the brightness of the spectrum matches the intensity of the adjacent individual interference patterns within the slit. From Tobin (2003), Figure 5.8.

Figure 2.2 shows the experimental device, which used Fresnel's mirrors to produce interference. These mirrors gave virtual images of a slit and thus played the same role as Young's slits, but with much higher brightness.

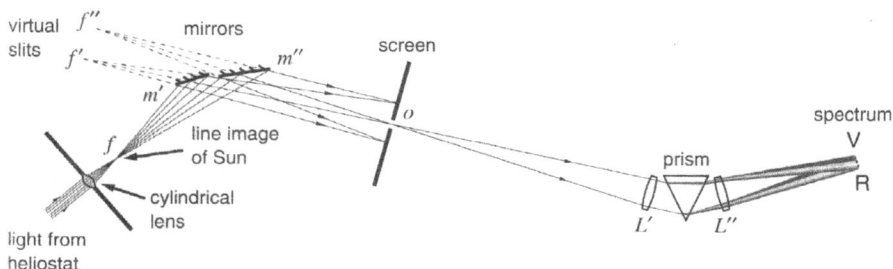

FIGURE 2.2 – Layout of the apparatus built by Fizeau and Foucault in 1845. The light source is the Sun, of which a linear image obtained by a cylindrical lens serves as an entrance slit. Fresnel's mirrors m' and m'' produce interference fringes on the screen, where a movable slit O is parallel to the central fringe. The spectroscope formed by a prism and the two lenses L' and L'' produces a striped spectrum as shown in Figures 2.1 and 2.3. The dimensions are not respected: there were about 2 meters between the mirrors and the screen, between the screen and the lens and between the prism and the spectrum. From Tobin (2003), Plate II.

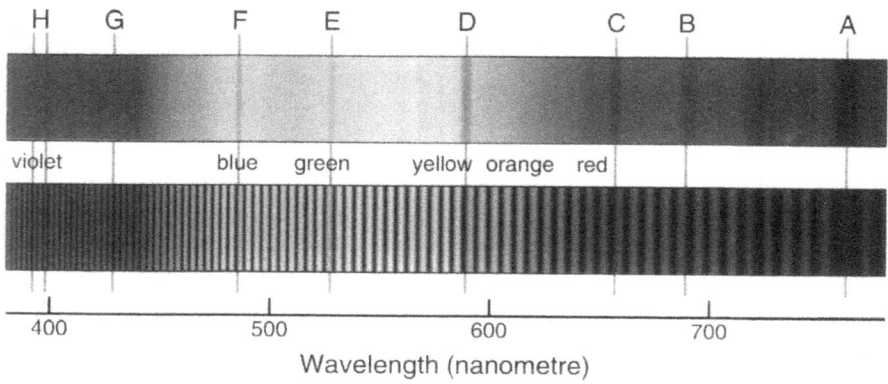

FIGURE 2.3 – The spectrum of solar radiation in visible light, with its main absorption lines lettered A-H by Fraunhofer (top), and observed with the apparatus of Figure 2.2, modulated by the stripes (bottom). From Tobin (2003), Plate III.

Figure 2.3 shows the actual appearance of the spectrum obtained with this device. The stripes modulate the spectrum of the Sun, allowing the determination of the path difference. Indeed, in this spectrum, each bright stripe at wavelength λ corresponds to an integer interference order n such that $\lambda n = $ constant (λn is then the path difference between the rays that interfere at this place). Considering stripes located approximately at the wavelengths λ_E and λ_F of the E and F lines of the solar spectrum, for example, the path difference of the incident rays corresponds respectively to $\lambda_E n_E$ or $\lambda_F n_F$. Knowing the two wavelengths λ_E and λ_F, it suffices to count the number of stripes between the E and F lines to determine the path difference. Conversely, once λn is known, one can determine any wavelength λ by counting the number of stripes that separates it from a line of known wavelength.

Fizeau and Foucault also mounted a variation of this experiment, where the striped spectrum was now produced using rotatory polarization in a crystal plate, a phenomenon discovered by Arago in 1811. The light from the Sun was linearly polarized by a Nicol prism, and then passed through a birefringent gypsum plate, which split it into two rays, the ordinary and extraordinary ray (Fig. 2.4). These rays interfered at the exit of the plate. As the refractive index in the gypsum was different for these two rays, their superimposition gave an elliptical polarization, with a variable aspect according to the phase difference between the rays (Fig. 2.5). This phase difference itself depended on the difference between the number of wavelengths of the ordinary ray and that of the extraordinary ray in the thickness of the plate. The phenomenon depended periodically on the wavelength, reproducing identical to itself when the path difference across the plate increased by one wavelength. The thickness of the plate was chosen to be sufficiently large for this to happen many times in the spectral range investigated. Next, an analyzer was placed in the beam, selecting a linearly polarized component. The spectrum of the light after crossing this analyzer was then modulated by regular stripes exactly as in the previous experiment, stripes that could be seen with a spectroscope. By rotating the analyzer by 90° one obtained bright stripes instead of dark ones. This set-up was much easier to install than the one of Figure 2.2, and was considerably more luminous.

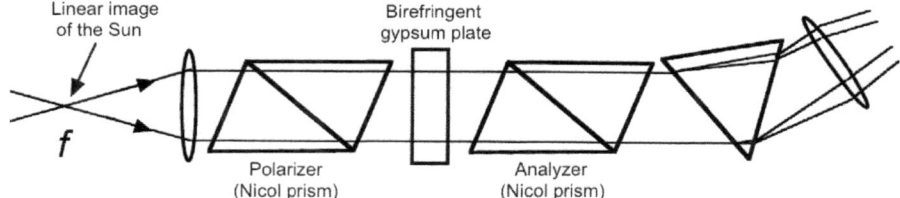

FIGURE 2.4 – The alternative experiment of Fizeau and Foucault. As in the first experiment (Fig. 2.2), a cylindrical lens formed a linear image of the Sun. A lens produced a parallel beam into which a Nicol polarizer and a birefringent gypsum plate were successively inserted then another Nicol prism as the analyzer. The light then fell onto a prism, which produced a striped spectrum similar to that of Figure 2.3. Author's drawing.

On the 24th of November 1845, Fizeau and Foucault presented their findings to the Academy of Sciences. They claimed to have obtained interferences with path differences of 813 and 1737 times the wavelength of the F line, i.e. respectively 0.4 and 1.4 millimeters. In spring 1846 they got fringes with a 3.2 mm path difference using rotatory polarization. All these results could be explained without difficulty using wave theory, and casted increasing doubt on the corpuscular theory, which was already much weakened. Our two scientists conclude:

"*The existence of these phenomena of mutual influence between two rays, in the case of large path differences, is interesting for the theory of light, in that it reveals in the emission of successive waves a persistent regularity that no phenomenon had indicated before.*"

This is what we presently call the coherence of light.

Despite the great interest of these experiments, which did not escape to the contemporaries, the work of Fizeau and Foucault was not to be published in the *Comptes rendus*, but only four years later in the *Annales de Chimie et de Physique*, a journal of which Arago was one of the editors[31]. We will see later that the two men returned to the nature of light in a completely different way, and that this was the occasion of a final quarrel, when their collaboration was transformed into competition.

---

[31] *Fizeau, H. & Foucault, L. (1849) *Annales de Chimie et de Physique* 26, 138-148; *Fizeau, H. & Foucault, L. (1849) *Annales de Chimie et de Physique* 30, 146-159.

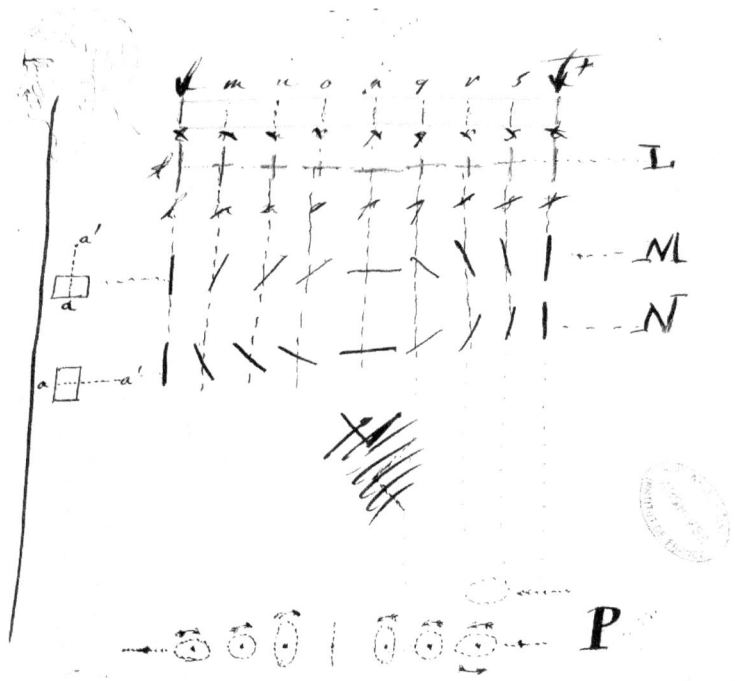

FIGURE 2.5 – A polarization cycle as a function of wavelength in the apparatus of Figure 2.4 (autograph drawing by Fizeau). In L, the length of the perpendicular lines indicated the change with wavelength of the intensity of the light when the analyzer of Figure 2.4 was either parallel or perpendicular to the polarizer. M and N show what happened for the two perpendicular positions of the birefringent gypsum plate indicated at the left. For M, the plane of symmetry aa' of the plate was parallel to the polarizer, so that only the ordinary ray entered the plate; the rotatory polarization turned the plane of polarization at the output of the plate as a function of wavelength, as shown on the figure. For N, the same occurred as when the plane of symmetry was perpendicular to the polarizer and only the extraordinary ray entered the plate. For P, the polarizer was at 45° from the plane of symmetry aa' of the plate; the drawings show the result of the interference between the ordinary and extraordinary rays, which both crossed the plate in this case: when the phase difference between these two rays varied between 0° and 180°, we went from a linear polarization to an elliptical polarization and then to a linear polarization perpendicular to the first one; then an elliptical polarization was observed anew but in the opposite direction when the phase difference varied from 180° to 360°. The analyzer selected a linearly polarized component of the light exiting the plate, whose full polarization state is described in P; the intensity of this component as a function of wavelength was maximum at the beginning and end of the cycle and zero in the middle if the analyzer selects the component drawn horizontally, as shown in L. This was the reverse if the analyzer selected the component drawn vertically. The repetition of this cycle produced a striped spectrum. Note the drawing of a head at top left. Such drawings are often found in the notes of the young Fizeau. Académie des sciences-Institut de France, Fonds 64.1 Hippolyte Fizeau, dossier 8.07.

2.2 The nature of the infrared radiations

In 1800, William Herschel (1738-1822) discovered the "radiant heat" of the Sun by placing, in the solar spectrum formed by a prism, small thermometers with a blackened bulb[32]. He noted that these thermometers were heated more to the red side of the spectrum than to the blue side, and were also heated further beyond the red, where the eye sees nothing. His later experiments showed that this radiation, whether from the Sun or from other sources such as a piece of red hot iron or even an oven ("the invisible culinary heat") was reflected by a mirror, as is light, and was more or less diffused or absorbed by various substances. But was that enough to ensure that this radiation was of the same nature as light? Herschel suspected it, but could not prove it. It is true that his attempt[33] to show the spectral energy distribution of light and of radiant heat (Fig. 2.6) did not encourage the readers to believe in this identity.

With the exception of Thomas Young[34], whose influence was unfortunately quite low, no contemporary and immediate successor of Herschel believed in that identity. For example, in his praise of Herschel before the Academy of Sciences, Joseph Fourier (1768-1830) in 1823 still spoke of "invisible rays of heat, mixed with sunlight." To complicate the problem, Johannes Wilhelm Ritter (1776-1810) in Germany and William Francis Wollaston (1731-1815) in England had independently discovered in 1801-1802 "invisible rays" or "chemical rays" that blackened a paper impregnated with silver chloride beyond the violet side of the solar spectrum.

---

[32] Herschel, W. (1800a) Investigations of the Powers of the Prismatic Colours to Heat and Illuminate Objects..., *Philosophical Transactions of the Royal Society of London* 90, 255-83, accessible via http://www.jstor.org/stable/107056
[33] Herschel, W. (1800b) Experiments on the Solar, and on the Terrestrial Rays that Occasion Heat..., *Philosophical Transactions of the Royal Society of London* 90, 437-538, accessible via http://www.jstor.org/stable/107062
[34] °Young, T. (1807) *A course of lectures on natural philosophy and the mechanical arts,* Vol. 1: new edition (1845), London, Taylor & Walton, p. 501-504.

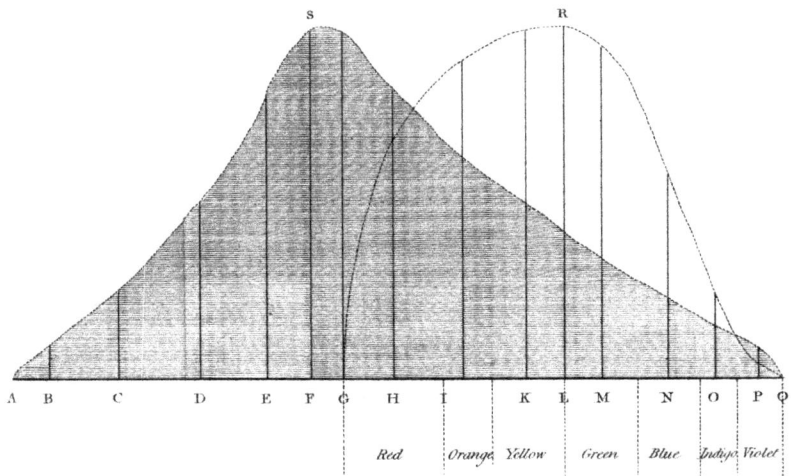

FIGURE 2.6 – William Herschel's attempt to show the spectral power distribution of light (R) and of the "radiant heat" (S) of the Sun. The distribution R of light, which is drawn schematically, mainly reflects the sensitivity of the eye. The radiant heat S was directly measured by a thermometer displaced in the spectrum. On a linear scale of wavelengths, such as that given by a diffraction grating, the distribution of the radiant heat would have its maximum at the yellow wavelength. However, the scale is far from linear here, as it is that of the dispersion of the prism, which varies as $1/\lambda^2$ with the wavelength λ: it is very dilated towards the violet side and contracted towards the red and infrared side with respect to a linear scale. Accordingly, the blackened bulb of Herschel's thermometer received energy within a range of wavelengths increasingly broader when moved from the violet to infrared, which displaced the observed maximum from yellow to deep red. Bibliothèque de l'Observatoire de Paris.

In 1835, André-Marie Ampère, a staunch defender of the wave theory of light, with his usual penetration offered a new idea, which is quoted at the beginning of an article[35] of Macedonio Melloni (1798-1854):

It consists in considering the radiant heat as a series of ripples in the ether excited by vibrations of hot bodies. These waves would be longer than the waves that make up light, if the heat source is obscure, but in the case of sources that are at the same time calorific and luminous, there would always be a group of waves simultaneously having the properties required to heat and illuminate. Thus, in this view, no essential difference exists between the radiating caloric and light.

---

[35] *Melloni, M. (1835) CRAS 1, p. 503-509.

It was still too early for this idea to be adopted: it did not convince Melloni, the leading infrared expert of the time, whom another expert, Samuel P. Langley (1834-1906), qualified in 1880 as "the Newton of heat." For Melloni, like most of his contemporaries, the radiant heat was the same thing as the *caloric*, this fluid studied by Fourier that seemed to spread slowly in bodies. For others, the minority, heat was an internal agitation of the body, which corresponds to our current knowledge and was progressively adopted, especially by Melloni. The latter finished in 1842 by accepting the identity of the nature of light and radiant heat, and even of "chemical rays", that is to say the ultraviolet[36]. He understood why the visible light spectrum schematized by Herschel and that of the radiant heat he measured looked so different from each other. He emphasized the importance of measuring the spectral distribution of radiation with an instrument only sensitive to energy, as was indeed more or less Herschel's thermometer. This is what we now call a *bolometer*.

Fizeau and Foucault were the ones who confirmed definitively the wave nature of infrared radiation by interference experiments. In the paper of 1847 in which they described these experiments[37], they wrote:

The many similarities revealed by experiments between the properties of heat rays and light rays, led to extend the idea of wave motions to heat rays. This view is widely accepted today, and yet it is only based on analogies, because none of the properties observed so far in the heat rays reveals in them a wave nature. The existence of interference phenomena would be decisive.

They first produced fringes with Fresnel's mirrors while isolating the infrared radiation through a dark red filter, and then displaced a thermometer with a narrow blackened bulb, manufactured by Fizeau itself, in these fringes: they observed the central fringe and two side fringes and saw fringes blur further because of the large breadth of the wavelength band. Then they performed Herschel's experiment again, this time with a good sensitivity and a much better spectral resolution. The result of this exploration of the solar spectrum with a thermometer is shown in Figure 2.7. Their device also allowed, as that of Figure 2.4, them to obtain a striped spectrum, allowing them to measure wavelengths in the

---

[36] *Melloni, M. (1842) *CRAS* 15, p. 454-460

[37] *Fizeau, H. & Foucault, L. (1847) *CRAS* 25, p. 447-450 and 485. Note that the authors used a permutation process between the source and surroundings and the surroundings themselves, a process that was to become the rule in infrared observations.

infrared for the first time, as we have explained above[38]. Fizeau alone did the calculations[39], because they were not the strong point of Foucault.

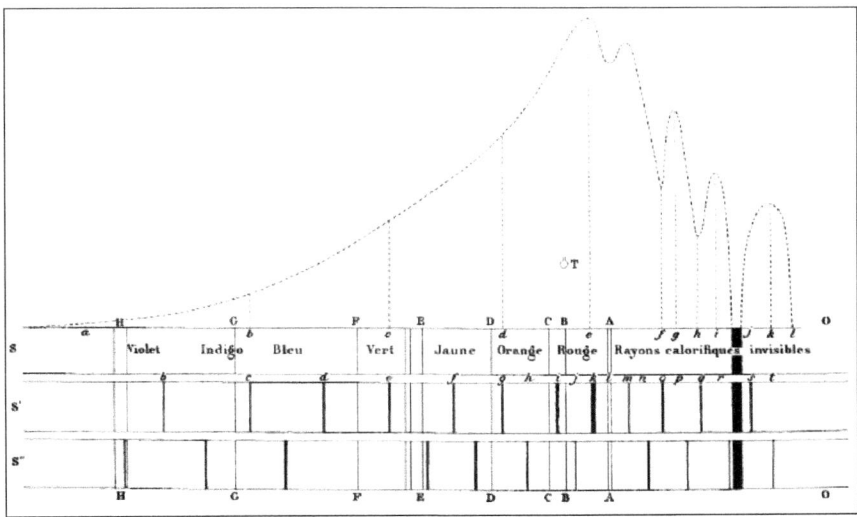

FIGURE 2.7 – The spectrum of the Sun obtained by Fizeau and Foucault. Top, the signal obtained by displacing a thermometer, whose diameter (about 1 mm) is indicated as **T,** in the spectrum. The strongest Fraunhofer lines are indicated by capital letters, but the resolution of the spectrum was insufficient to see them. However, the infrared absorption bands of the atmospheric water vapor, to the right, are clearly visible. They had already been identified by John Herschel (1792-1871), the son of William, who was himself an astronomer and physicist of high quality. In **S'**, the lower case letters locate the position of the spectral stripes provided by a device similar to that of Figure 2.4. In **S"**, the analyzer was rotated by 90° and the stripes were inverted, the bright ones taking the place of the dark ones of **S'** and vice versa (see legend of Fig. 2.5). These stripes allowed measurement of wavelengths in the infrared for the first time. From Foucault (1878), Bibliothèque de l'Observatoire de Paris.

Fizeau and Foucault also observed the diffraction of heat rays by the edge of a screen. The text in which they presented their work is of great modesty, despite the inventiveness and the quality of their experiments. This modesty is probably the reason why they appear to have been somewhat forgotten thereafter: in the second edition dated 1868 of

---

[38] With this set-up, we have to assume that the refraction indices of both the ordinary and the extraordinary rays in the birefringent crystal do not vary with wavelength. The set-up of Figure 2.2 does not have this inconvenience.

[39] °Fizeau, H. (1847) *Procès-verbaux de la Société philomatique de Paris*, p. 108-109.

his famous *Cours de physique de l'École polytechnique*[40], Jules Celestin Jamin (1818-1886), who seems to refer to Melloni, did not mention them and still spoke of *"the probable identity of heat and light."* Yet no one should have disputed that identity after these wonderful experiments.

Fizeau and Foucault parted ways and worked alone, until an unfortunate competition was to bring them together again and instigate their final quarrel.

---

[40] °Jamin, J. (1868) *Cours de physique de l'École polytechnique*, 2ᵉ ed., Paris, Gauthier-Villars, p. 263.

Chapter 3
The Doppler-Fizeau effect

The first experimental verification of the Doppler-Fizeau effect took place in 1844 using a train travelling between Utrecht and Maarssen in the Netherlands. The locomotive should have looked like this one, a replica of a machine built by Arend, Longridge & Co in England. The track gauge was 1.945 m, wider than the current standard gauge of 1.435 m. Railway museum, Utrecht, Wikimedia Commons, Quistnix.

3.1 Christian Doppler

In 1842, Christian Doppler (Fig. 3.1, Box 3.1), professor of mathematics and practical geometry at the Polytechnic school (also called Technical Institute) in Prague, published, in the Proceedings of the Royal Scientific Society of Bohemia, the text of a lecture given before that society on 25 May[41]. The translation of the full title is: "On the colored light of double stars and some other stars of the sky; Test of a general theory that incorporates Bradley's theorem on the aberration as an integral part."[42] This title does not explicitly state what is, for us, the whole point of the article: the demonstration that the observed wavelength of sound, or that of light, depends on the relative velocity of the source and the observer.

FIGURE 3.1 – Christian Doppler (1803-1853), near 1836, unknown artist. Austrian Academy of Sciences, Vienna. Wikimedia Commons, Kelson.

---

[41] Doppler, C. (1842), *Abhandlungen der königliche böhmischen Gesellschaft des Wissenschaften*, 2, p. 465-482. The original text in German is accessible via http://de.wikisource.org/wiki/Über_das_farbige_Licht_der_Doppelsterne_und_einiger_anderer_Gestirne_des_Himmels. A commented summary in English is in http://en.wikipedia.org/wiki/Über_das_farbige_Licht_der_Doppelsterne_und_einiger_anderer_Gestirne_des_Himmels. A fac-simile reproduction of the text and a full translation into English can be found in Eden (1992).

[42] In reality the work has little to do with the aberration discovered by Bradley.

Box 3.1. Christian Doppler

Christian Doppler was born in Salzburg on the 19th of November 1803, into a family of marble cutters. His mathematical gifts allowed him to enter the Vienna Polytechnic Institute, where he studied from 1822 to 1825. In 1829, he became assistant in this institution, and tried to find a permanent job. After some difficult years, he obtained a chair in a secondary technical school in Prague in 1835. His teaching schedule was very full and left him little time for personal research. In 1841 he was appointed professor at the Polytechnic Institute of Prague. Subjected to slander, he resigned in 1844. Supported by the philosopher and mathematician Bernard Bolzano, he resumed his position in 1846, and obtained a professorship of mathematics, physics and mechanics the following year at the Mining and Forestry Academy in Banská Štiavnica, in today's Slovak Republic. The consecration came in 1848 with his election to the Imperial Academy of Sciences in Vienna and his appointment in 1850 as director of the Institute of Physics of the University of this city. But he had been sick since 1844, probably of tuberculosis, from which he died on the 17th of March 1853 in Venice, to where he had travelled hoping for an improvement. The scientific work of Doppler is very varied, ranging from pure and applied mathematics to astronomy and optics. But it is his 1842 paper which ensured his posthumous fame.

After an introduction recalling that the wave nature of light was not yet fully adopted by all, let alone the transverse nature of its vibrations, Doppler noted:

"One has been stubborn in not considering the light or sound waves, when they cause the sensation of light and sound, as objective phenomena: one wandered after what period of time and with what intensity these undulations themselves were produced, without asking after how much time and with what subjective intensity these ripples were perceived by the eye or the ear of the observer. However, it is this subjective impression [of the sound or light received by the observer], and not the objective nature of the ripples [of the sound or light emitted by the source], that determine the color and its intensity, or the tone and the strength of the sound. [...]

For example, as long as the observer and the source of the waves keep their relative positions, one cannot doubt that, anything else being unchanged, the subjective estimates agree numerically with the objective conditions. If, however, the observer and the source of the waves change place individually or together, if they move away or closer with a speed of the same order as the velocity of propagation of the wave, what will happen? [...] Is it not obvious that the time interval between two waves will be shortened if the observer approaches the origin of the wave motion? Will it not increase, if he moves away? In the first case, the pitch and intensity of the wave should increase; they should decrease in the second case. A motion of the source of the waves should produce similar modifications. Everyday's experience shows that a ship with a fairly large draft is beaten by more waves and much more shaken if she governs against the waves, that if she is carried away by them, or as if she is at rest. Should not we find in the sound and light waves what occurs in fluid waves? Some simple formulae will enlighten this question."

Doppler therefore sought to develop the phenomenon that he just described in equations, using small sketches. He only treated the simplest case where the movement is collinear with the propagation of the wave.

Box 3.2. The Doppler-Fizeau effect.

Let us consider a source of sinusoidal sound or a light wave with a period T (or if we prefer a frequency $\nu = 1/T$) and with a wavelength $\lambda = cT = c/\nu$, c being the velocity of propagation of the wave. Let us suppose that the source approaches the observer with the velocity u, and consider the received wave at a point A located in the direction of motion. A maximum of the wave is received in A after a transit time t. The next wave maximum would have been received after the time $t + T$ if the source had not moved. But as it moved by uT between the two maxima of transmission, the second maximum is received in A after a time $t + T - uT/c = t + (1 - u/c)T$. The apparent period of the wave received by the observer at A is thus not T but is reduced to $(1 - u/c)T$. In contrast, the apparent period of the wave will increase as $(1 + u/c)T$ if the source moves away from the observer with the velocity u.

If we now consider a fixed source and an observer who moves, we must distinguish the case of sound and the case of light.

The Doppler-Fizeau effect

In the case of sound, there is no symmetry between the two situations: moving source or moving observer, because there is a medium in which the sound propagates. Indeed, if the source is moving away from a stationary observer, the latter will always receive a wave even if $u > c$, while if the observer moves away from a fixed source with a speed greater than c he will receive nothing. In the case of a moving observer approaching a stationary source, let us place ourselves in the frame of the medium where the sound propagates: during the time T' between the receipt of two successive maxima, the wave has advanced by cT' and the observer by uT'; the sum $(u+v)T'$ is equal to the wavelength λ, i.e. $(u+v)T' = \lambda = cT$. The apparent period of the wave is $T(1+u/c)^{-1}$.

In the case of light, an identical result is obtained whether the source or the observer is fixed, because, according to the principle of relativity, only the relative movement counts. If the source and the observer approach each other with the velocity u: the period is reduced to $T' = (1-u/c)T$; the frequency is increased to $\nu' = \nu(1-u/c)^{-1}$ and the wavelength is reduced to $\lambda' = (1-u/c)\lambda$ (blue shift). If the source and the observer move away from one another with the velocity u: the period is increased to $T' = (1+u/c)T$; the frequency is decreased to $\nu' = \nu(1+u/c)^{-1}$ and the wavelength is increased to $\lambda' = (1+u/c)\lambda$ (redshift).

If the velocity u is close to the speed of light c, the period and the wavelength given by the above expressions should be divided by $(1-u^2/c^2)^{1/2}$, the frequency being multiplied by the same quantity. In the case of sound, the above expressions are valid if it is the source that moves relative to the medium in which the propagation takes place, where the stationary observer is located. But if the observer moves relative to the source and the medium, he observes the frequency $\nu' = \nu(1+u/c)$ in the case of approach, and $\nu' = \nu(1-u/c)$ in the opposite case.

Doppler gave formulae for the received frequency ν', the emitted frequency being ν. In modern notations, they are:
$\nu' = \nu(1-u/c)^{-1}$ if the source is approaching the observer with the velocity u;
$\nu' = \nu(1+u/c)$ is the observer approaches the source with this velocity.

Doppler then explicitly calculated the speed at which the observer must move towards a sound source emitting the note C_3 to hear D_3 and found it is 128 Parisian feet (32.48 cm) per second, while, if it is the source that approaches, this is 114 feet per second.[43]

However, Doppler's main aim was to explain the color of the stars by their radial velocity[44], i.e. the speed with which they move away or closer to us. He claimed that a relatively low speed of about one hundred kilometers per second (in modern units) was sufficient to cause a change of color perceptible to the human eye. Moreover, apparently unaware that the spectra of stars extends beyond the limits of the sensitivity of the eye, he thought that the stars would disappear completely from view if their radial velocity exceeded 136,000 km/s in either direction. He applied these ideas to account for the variety of colors of double stars: according to him, isolated stars are white, but the components of double stars that revolve around each other with a speed that he considered to be very large, could become red if they move away or blue if they approach. He also explained the variability of certain stars, that would be double star components that become visible when their radial velocity is appropriate. In particular he explained, by variations in velocity, the appearance and the gradual disappearance of the supernovae of Tycho Brahe (1572) and Kepler (1604), and the possible change of color of Sirius, which was red in Antiquity and was now "dazzling white".[45]

Doppler concluded:

"The purpose of the present Memoir was to highlight, not the accidental possibility of this connection [between the color and radial velocity of stars], but its essential necessity; and it is not a slight satisfaction for the author to be able to assert that the theory agrees fully with observations."

Needless to say, there is nothing left of these conclusions: the orbital velocities of the components of double stars are much too low, at most a few hundred kilometers per second, to affect their color, which is an intrinsic property related to their temperature. However, there are very distant galaxies with a very high recession velocity, whose redshift is

---

[43] These results are exact with a velocity of sound of 340 m/s, and an equal temperament.
[44] The term of radial velocity (*vitesse radiale* in French) seems to have been used for the first time by Deslandres: see *Deslandres, H. (1891) *CRAS* 113, p. 737.
[45] For the change of color of Sirius and possible explanations, see +Bonnet-Bidaud, J.-M. & Gry, C. (1991) The stellar field in the vicinity of Sirius and the color enigma, *Astronomy & Astrophysics* 252, 193-197.

such that their emission in the far-ultraviolet is observed in the red and even in the infrared.

Doppler gave no experimental verification of the theory, which did not generate much enthusiasm outside a small circle. A colleague of Doppler in Prague, Karl Kreil (1796-1862), seems to have been the first to discuss and accept the principle; he suggested that we could check the color changes of the stars predicted by Doppler by dispersing their light with a prism. But he did not mention the absorption lines described in 1814 by Fraunhofer in the spectrum of the Sun and Sirius, lines that should be displaced as well as colors. The famous Prague mathematician and philosopher, Bernard Bolzano (1781-1848), an admirer of Doppler, was also interested in the principle while expressing some doubts about his explanation of the color of double stars, believing that if it were true we should observe the color changes in their orbital motion[46]. But, above all, a Dutch physicist, Hendrik Christophorus Buijs Ballot (Fig. 3.2), also spelled Buys Ballot (1817-1890), better known as the founder of the Dutch meteorological, took Doppler's conclusions as the subject of his thesis in 1844.

3.2 Christophorus Buijs-Ballot

FIGURE 3.2 – Christophorus Buijs-Ballot (1817-1890). Wikimedia Commons.

A summary in German of the thesis of Buijs Ballot[47], whose title can be translated as "Acoustic experiments on the railway of the Netherlands, with a few remarks on the theory of Professor Doppler",

---

[46] °Bolzano, B. (1843), *Annalen der Physik und Chemie* 2^d ser., 60, p. 83-88. An abridged translation into French is given by °Moigno (1850) p. 1182-1184.

[47] °Buijs-Ballot, C.H. (1845) *Annalen der Physik und Chemie,* 2^d ser. 66, p. 321-351. There is an abridged translation into French in °Moigno (1950), p. 1185-1189.

contains in the front page a sentence in Latin that we translate as follows: "Mr. Doppler's theory needs to be checked, but I say that it is not sufficient to explain the colors of double stars." We must realize that while the acoustic Doppler effect is readily observed today, thanks to the sirens of firefighters or police cars and ambulances, only the trains were fast enough at the time to observe it. Thus, it is on trains that Buijs Ballot performed his experiments, as described as follows by Moigno[48]:

"On the UM railway from Utrecht to Maarssen, three groups of musician observers were placed at distances nearly equal to one thousand meters. They remained motionless, as close to the track as possible. A musician placed on the locomotive sounded the trumpet, first starting from Utrecht, then between A and B, between B and C, and between C and Maarssen. Mr. Buijs Ballot published in two tables estimates of the pitch of the received sound, made in two series of experiments on 3 and 5 June 1845 and concluded that the numbers obtained were in general agreement with the theory."

Moigno added:

"After reporting on his experiences, the first of their kind, Mr. Buijs Ballot poses the following five questions and discusses them:
1. Have we the right to extend to light the results obtained with the sound?
2. Have stars, at any point in their orbit, a velocity sufficient to cause a significant coloration or a visible color change?
3. Do double stars have intrinsic colors, or do they actually change colors as assigned by Mr. Doppler?
4. Is there an easier explanation of the color of the stars?
5. Are there any facts that make it impossible to apply to the color of the stars M. Doppler's theory?"

Buys Ballot answered positively to the first question and negatively to the second one, based on various authorities. He was less confident regarding the third question, while taking over the objections of Bolzano. And finally, in response to the fourth question, he wrote that "the differences in color of the stars are due to the nature of the light of these stars, and not to their movements. [...] Fraunhofer's experiments demonstrate the essential differences in their light, showing that the lines of the spectra obtained with the light of the various stars are actually different in their positions, their extent [width], their brightness, etc."

---

[48] °Moigno (1850), p. 1185.

Indeed, in 1814 Fraunhofer had been able to obtain not only the spectrum of the Sun, in which he saw the numerous absorption lines to which his name has been given, but the spectrum of the brightest star in the sky, Sirius. He noticed that its lines are very different from those of the Sun, and that there was an absorption band at 430 nm that he called G, hence the name given to the spectrum of this type of star. Nine years later, he observed other stars with still different spectra, and recognized a strong solar line in the spectrum of Sirius (now known to be due to hydrogen, but Fraunhofer could not know that). It is because of this confused situation that Doppler did not consider to use the spectral lines to measure the velocity of stars, but only their color, which was a total failure.

Doppler could not leave Buijs-Ballot's criticism unanswered. The title of his new article[49] of 1846 can be translated as "Notes on my theory of the colored light of double stars, with some observations on the objections of Mr Buijs-Ballot, etc." He first expressed his satisfaction with the result of the acoustic experience of the latter, which confirmed his theory, then embroiled in weak and confused explanations that it would take too long to detail. He remained convinced "that the objections raised by no means overturn [his] theory, that time will avenge the premature attacks of which it was the object, and that it is a great way to penetrate further into the depths of the heavens." If he was right on the last point, he did not convince anyone of the validity of his explanation of the color of the stars.

3.3 Hippolyte Fizeau

Fizeau certainly ignored the work of Doppler and the following discussions. The proximity of the date of the communication[50] that he gave to the *Société philomathique*, 23rd of December 1848, with that of the controversies published in German in Poggendorf's *Annalen der Physik und Chemie*, is purely coincidental. Fizeau, who was very honest, did not cite Doppler and Buijs-Ballot, and his approach was quite

---

[49] *Doppler, C. (1846) *Annalen der Physik und Chemie* 2^d ser. 68, p. 1-35. French summary in °Moigno (1850), p. 1189-1195.
[50] °Fizeau, H. (1848) *Procès-verbaux de la Société Philomathique de Paris*, p. 81-83. This is a summary, also reproduced in °Moigno (1850), p. 1197-1199, who gives an image of the demonstration apparatus of Fizeau, and adds comments. The full paper was only published in 1870: °Fizeau, H. (1870) *Annales de Chimie et de Physique*, 4th Ser., 19, p. 211-21. The citations in our text come from the summary.

different. Besides, he did not consider an experiment with the sound from a train, but wondered "what would perceive a man [walking] as a fast soldier."[51] He first discussed the sound waves, and wrote:

> *"If a source emitting a continuous and constant sound moves with a speed comparable to that of sound, the sound waves will not be symmetrically arranged around the source, as occurs when it is at rest; but they will be closer to each other in the region towards which the movement takes place and more separated in the opposite region; for an observer placed in front or behind the source, the sound will be different, with a higher pitch in the first position, and a lower one in the second.*
> *If the observer is supposed to be in motion, the source of the sound being fixed, the result will be similar; but the law of the phenomenon is different.*
> *Calculating the velocities that correspond to the intervals of the musical scale, we find the following numbers: to produce an elevation by a semitone, the source must have a speed of 21.25 [m] per second, for a major tone 37.8, for a third 68, and for an octave 170 [This implies a sound velocity of 340 m/s]. In the case of a stationary source, and for the same notes the observer must have respectively the speeds: 22.6; 42.5; 85; and 340. The sounds emitted or received in different directions from those of the motion are calculated by projecting the velocity on the same direction."*

To illustrate his principle, Fizeau brought a simple and very clever demonstration apparatus to the *Société philomathique* on the 23rd of December 1848, shown in Figure 3.3. He described it as follows:

> *"This device is based on the principle of the toothed wheels of M. Savart, but the disposition is inverted. Instead of movable teeth encountering a fixed elastic body in their motion, the elastic body [a flexible blade] is placed on the circumference of a wheel and meets fixed teeth arranged on the concavity of a stationary outer arc in its movement. It is thus a fixed device that has the property of emitting different sounds in each particular direction."*

We recognize here the interest of Fizeau for toothed wheels, which soon led to the device with which he measured the speed of light.

---

[51] Académie des sciences-Institut de France, Fonds 64 J Hippolyte Fizeau, dossier 8.10.

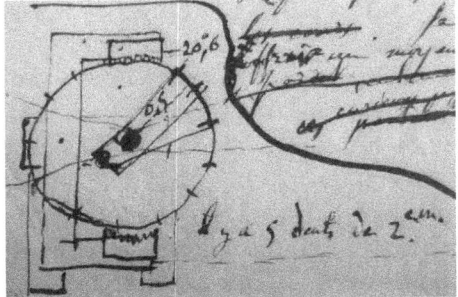 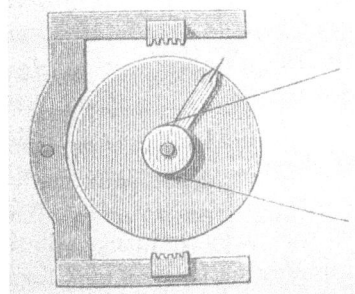

FIGURE 3.3 – Fizeau's apparatus to demonstrate the acoustic Doppler-Fizeau effect. Left, a drawing by Fizeau in 1848 (Archives de l'Académie des sciences, Fonds 64.1, dossier 8.10), right, from °Fizeau, *Annales de Chimie et de Physique, 4th Ser.*, 19, p. 213 (Bibliothèque de l'Observatoire de Paris). The arm, rotated by a crank, carried at its end a flexible blade which met the fixed rack in its movement, producing a sound like a rattle. The wheel had a diameter of 1 m. Assuming that the arm rotated in the clockwise direction, the sound produced by this mobile source appeared lower to an observer on the left who heard the sound of the blade rubbing on the upper rack and moving away from him, than to an observer on the right. This was the reverse when the blade was rubbing the lower rack. A distant observer on the left or right heard a double sound.

The *Société philomathique*, antechamber of the Academy of Sciences, certainly enjoyed the demonstration of Fizeau, because he was elected as a member on the following 27th of January.

But here is the main point of the article of Fizeau, where one finds his real novelty compared to Doppler: he crossed the big step between the vague assimilation of light with sound and the accurate description of the phenomenon created by the motion of a light source:

"A very quick movement comparable to the speed of light, assigned to a luminous body or to the observer, has the effect of altering the wavelength of all the simple rays that make up the light received in the direction of motion. This wavelength is increased or decreased depending on the direction of the motion. Considered in the spectrum, this effect will result in a <u>shift of the lines</u> [emphasized by Fizeau] corresponding to the change in the length of the undulations."

Then Fizeau considered measuring the radial velocity of Venus during its orbital motion: he calculated the value of the change in deviation, for the sodium D line, of the light passing through a flint prism (a highly refractive lead glass) of 60° angle, when Venus was moving away with its maximum speed of 35 km/s; he determined a value of 2".65. For the orbital velocity of the Earth (30 km/s) this was to be 2".25. He believed that this would be detectable, especially by placing two prisms in tandem and observing successively at times when the movements were in the opposite directions. He thought that "the difficulties are not such that we cannot hope to overcome them." But he does not seem to have done this experiment.

Taking cognizance of Fizeau's work in 1850, Doppler welcomed the confirmation of his theory by the acoustic demonstration apparatus, and claimed that "on the acoustic point of view, my theory therefore can now be considered as almost put out of doubt, but not to the same degree in optics."[52] Indeed, he did not see how it could be confirmed by experiment in the case of light. And he did not seem to have grasped the scope of the idea of displacement of spectral lines introduced by Fizeau. He did not say a word about it and came back to its colored stars, to which he even returned once more two years later[53]. His attitude is not surprising, for the consideration of the spectral lines was still not widespread, and the diversity of stellar spectra recorded by Fraunhofer was confusing the minds. It is the deep knowledge of spectral analysis that Fizeau had acquired in his work with Foucault that allowed him to correctly interpret the influence of the motion of a light source, avoiding the sterile way to which Doppler was committed.

To summarize, Doppler has shown how the relative movement between a source of sound and an observer changes the apparent period of the vibrations he receives. But it is Fizeau who first saw clearly how this principle can be applied to the propagation of light waves, and how to measure the relative velocity of the source and the observer in this case, using spectral lines. It is therefore legitimate to associate their names, considering the *Doppler-Fizeau effect*. However, in the late nineteenth century already, one no longer spoke of Doppler-Fizeau, only of Doppler[54].

3.4 The future of the Doppler-Fizeau effect

Even if it was independently discovered by Doppler and Fizeau, their effect hardly seemed likely to lead to practical applications at the time. However, if measuring the displacement of stellar spectral lines seemed difficult, scientists did not forget this possibility as in the following years we find many references of this in their publications.

---

[52] *Doppler, C. (1850) *Annalen der Physik und Chemie* 2^d ser. 81, p. 270-275.
[53] *Doppler, C. (1852) *Annalen der Physik und Chemie* 2^d ser. 85, p. 371-378.
[54] See e.g. *Faye, H. (1891) *CRAS* 112, p. 281.

The discovery by Gustav Kirchhoff and Robert Bunsen of the presence of terrestrial elements in the Sun, made in 1860 thanks to spectroscopy, reactivated the interest of physicists and astronomers for spectral analysis. At the same time, Father Angelo Secchi obtained over 4,000 spectra of stars and other objects at the Vatican observatory, and posed, with others, the bases of stellar classification. He hoped to see the spectral shift due to the radial velocity of various stars, but without success. In 1863, Sir William Huggins (1824-1910, Fig. 3.4) also obtained spectra of stars and nebulae and tried, for the first time, to measure the radial velocity of Sirius and Aldebaran: this was unsuccessful because his spectroscope did not have sufficient dispersion.

FIGURE 3.4 – Sir William Huggins (1824-1910). Bibliothèque de l'Observatoire de Paris.

Huggins then built the more efficient spectroscope shown in Figure 3.5, and succeeded in seeing the displacement of a spectral line of Sirius for the first time in 1868 (Fig. 3.6) and in measuring the radial velocity of the star. He found that Sirius was moving away from the Earth at a speed of 66 km/s[55]. After correction for the orbital motion of the Earth, this gives a radial velocity relative to the Sun of 47 km/s. The current value is –7.6 km/s, corresponding to a velocity relative to the Earth of 11 km/s at the time of observation! So the measurement

---

[55] Huggins, W. (1868) Further Observations on the Spectra of Some of the Stars and Nebulae, with an Attempt to Determine Therefrom Whether These Bodies are Moving towards or from the Earth..., *Philosophical Transactions of the Royal Society of London* 158, p. 529-564, accessible via http://www.jstor.org/stable/108925. Huggins wrote erroneously p. 549 that the radial velocity of Sirius of 29.4 miles per second (47 km/s) that he derived from his observations was relative to the Earth; actually, this is the velocity relative to the Sun, as can be seen from the context. This error propagated to Cornu, A. (1890).

was wrong. Huggins was well aware of the uncertainty of his result, which comes from the difficulty of alignment of the stellar spectrum with the comparison one, and more generally of visual observation. For his part, Secchi did not find a displacement of the lines of Sirius[56].

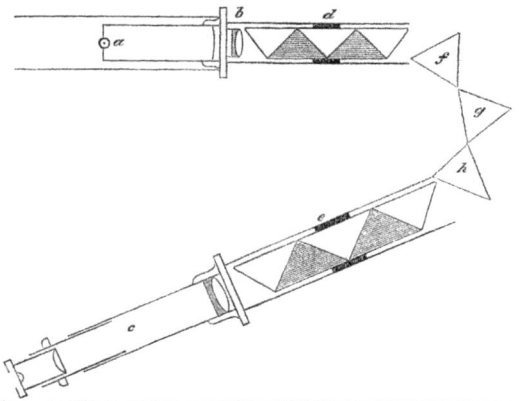

FIGURE 3.5 – The spectroscope of Huggins (1868). The light reached the entrance slit *a*; the lens *b* gave a parallel beam which passed through the train of prisms *d*, whose mean deviation was zero but which nevertheless strongly dispersed light, then through the prisms *f*, *g* and *h*, and finally the second train of prisms *e*. The resulting spectrum was examined with the small telescope *c*. The prisms in grey, were made of flint, a glass with high dispersive power, the others were in crown glass. Bibliothèque de l'Observatoire de Paris.

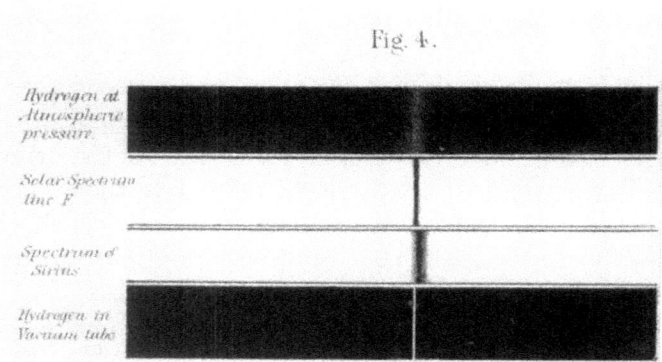

FIGURE 3.6 – The F line of hydrogen at 487 nm, visually observed by Huggins in Sirius. Huggins also observed it in comparison spectra of the Sun and of hydrogen at various pressures. Bibliothèque de l'Observatoire de Paris.

---

[56] *Secchi, A. (1868) *CRAS* 66, p. 398-403.

It is remarkable that, at the time of Huggins' measurement, the Doppler-Fizeau effect was neither well known nor, when it was known, accepted by everyone, at least as far as light was concerned. Huggins had to appeal to the authority of John Clerk Maxwell (1831-1879) to support his claim, citing in his article of the *Philosophical Transactions* a long letter from the latter. There, Maxwell gave again the formulae for the effect[57]. Maxwell apparently did not know the 1848 paper of Fizeau: he quoted Fizeau, but for two other articles. Henri Poincaré (1854-1912) himself had questions about the Doppler-Fizeau effect and encouraged astronomers to make verifications. One of them was to take the spectrum of the Sun with a long slit placed along its equator: the rotation of the Sun produces a difference of velocity by 4 km/s between the two ends. The first convincing measurement of this effect appears to be that of an American astronomer, Charles Augustus Young (1834-1908), who in 1876 obtained a very high dispersion spectrum using a diffraction grating[58]. Another measurement was made in 1880 by Louis Thollon (1829-1887), who used a high dispersion spectroscope with a prism of carbon disulfide at the Observatory of Nice: he compared for the two edges of the Sun, the solar lines with telluric lines (due to absorption by the constituents of the atmosphere, and whose wavelength is fixed), and observed the difference in their velocities[59]. Thollon also measured the radial velocity of protuberances at the edge of the Sun with his spectroscope, a result that proved difficult to convince some of his contemporaries of.

Another observation was desired by Henri Poincaré to solve a problem with the Doppler-Fizeau effect: what is the observed radial velocity for a light source that does not emit by itself, but by reflection or diffusion? Is it actually that of the source? Poincaré showed in 1895 that a planet illuminated by the Sun returns to Earth by scattering the light of the Sun with an apparent radial velocity which is the algebraic sum of the velocity of this planet relative to the Sun and of its velocity

---

[57] Maxwell added a small correction whose origin is unclear, which might correspond to the composition of the velocity of light with that of the Earth. In any case, such a term does not exist because the velocity of light is a constant.

[58] +Young, C.A. (1876) Observations of the displacement of lines in the solar spectrum caused by the Sun's rotation, *Memorie della Societa Degli Spettroscopisti Italiani* 5, p. A143-51.

[59] See *Cornu, A. (1890), p. D.25.

relative to the Earth.[60] But he wanted to have confirmation by observation. Deslandres realized the wish of Poincaré by placing the slit of a spectrograph along Jupiter's equator (Deslandres used photography to record his spectra): he observed the overall displacement of the spectrum predicted by Poincaré and also the inclination of the lines due to the rotation of the planet[61]. A simpler case, that is familiar to us, is the measurement of the speed of a car by radar: in this case, the radiation source is fixed relative to the receiver and the displacement of the frequency gives directly twice the speed of the vehicle.

Deslandres also confirmed an observation made in 1895 by James Keeler, who had measured the rotation of Saturn's rings by spectroscopy[62] (Fig. 3.7), showing that they did not rotate as a solid body, in agreement with the theoretical prediction of Maxwell.

Poincaré's interest in the Doppler-Fizeau effect was clearly related to the problems of propagation of light and of reference systems that were at the center of the preoccupations of the physicists of the time: they were to lead to Einstein's special relativity in 1905.

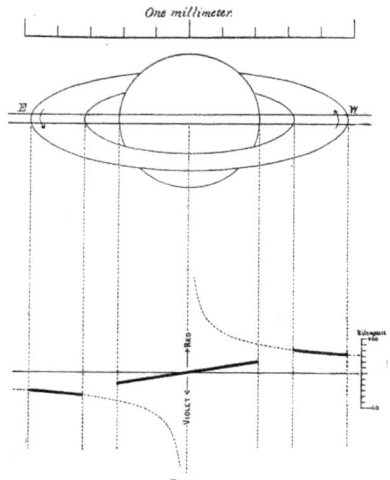

FIGURE 3.7 – The shape of a line of Saturns spectrum, observed with a long slit by Keeler. Top, the position of the slit on Saturn and its rings. Bottom, aspect of the line distorted by the Doppler-Fizeau effect: the planet rotates as a solid body, while the particles forming the rings orbit the planet not as a solid body, but according to Kepler's laws (the dotted line corresponds to the theoretical prediction). From +Keeler, J. (1895) *Astrophysical Journal* 1, p. 416, with thanks to the American Astronomical Society.

---

[60] *Poincaré, H. (1895) *CRAS* 120, p. 420-421. The result had already been demonstrated by +Niven, C. (1874) On a method of finding the parallax of double stars and on the displacement of the lines in the spectrum of a planet, *Monthly Notices of the Royal Astronomical Society* 34, p. 339-347.

[61] *Deslandres, H. (1895) *CRAS* 120, p. 417-420.

[62] *Deslandres, H. (1895) *CRAS* 120, p. 1155-1158; +Keeler, J. (1895) A spectroscopic proof of the meteoritic constitution of Saturn's rings, *Astrophysical Journal* 1, 416-427, Plate VIII.

It is not surprising either that several scientists tried to check the Doppler-Fizeau effect on light in the laboratory, as Buijs-Ballot had done for sound. Two attempts were made in St Petersburg with fairly complex devices but with only marginal results.[63] In 1914, Charles Fabry (1867-1945) and Henri Buisson (1873-1944) gave a perfectly conclusive demonstration with a very simple device, which is the equivalent of what Fizeau had used for sound waves (see Fig. 3.3).[64] A white paper disk mounted horizontally on a manual centrifuge rotated at high speed. It was illuminated by a monochromatic mercury lamp, and when either side of the disc was observed almost tangentially with a Fabry-Perot interferometer, the interference rings were seen to contract or to expand when the disc rotated, due to the variation of the wavelength. The effect was measured to within 2% without special care. The authors stated that, if more precautions were taken, this could have been a good method for measuring the speed of light.

To return to the stars, several observatories started systematic programs of observation of their radial velocity: Greenwich in 1875, Potsdam in 1888, Paris in 1891, etc. Comparison of the results obtained on the same stars shows that errors were still very large, often reaching several tens of km/s. Photographic spectrographs improved the accuracy, while giving access to the blue and ultraviolet part of the spectrum. But the number of stars observed remained small for a long time, so that the radial velocities were not actually used in studies of stellar kinematics before the second half of the twentieth century.

What was Fizeau doing on the subject while the radial velocity measurements were developing? Not much, because his research focused on other things and also because he was overloaded with tasks given to him by the Academy of Sciences. Nevertheless, the fact that he decided to publish his former communication to the *Société philomatique* in 1870 was a way of reminding everyone of his priority after the publication of

---

[63] +Belopolsky, A. (1901) On an apparatus for the Laboratory Demonstration of the Doppler-Fizeau Principle, *Astrophysical Journal* 13, p. 15-24; +Galitzin, B. & Wilip, J. (1907) Experimental test of Doppler's Principle for Light-Rays, *Astrophysical Journal* 26, p. 46-58. The insertion of these papers in an American journal testifies for the interest of the experimental verification of the Doppler-Fizeau effect for light. Belopolsky had received 300 dollars from the Elizabeth Thompson Science Fund to perfom his experiment.
[64] *Fabry, Ch. & Buisson, H. (1914) *CRAS* 158, p. 1498-9.

the article of Huggins[65], and also of Father Secchi who believed to have seen the Doppler-Fizeau effect due to the rotation of the Sun that year[66].

The first systematic measurements of the radial velocities of galaxies were performed in 1912 by Vesto M. Slipher (1875-1969)[67]. They required 20 to 40 hours of exposure with the 61-cm diameter telescope of the Lowell Observatory in Flagstaff (Arizona). These exposure times seem extraordinarily high to us today. It is well known that from 1927 to 1929 Georges Lemaître (1894-1966) and Edwin Hubble (1889-1953), who had many radial velocity measurements of galaxies at their disposal, found the proportionality between the radial velocity and the distance of galaxies due to the expansion of the Universe independently of each other[68].

It would be futile to attempt to enumerate the many current applications of the Doppler-Fizeau effect, which range from astronomy to physics and medicine, and extend to daily life. Among the most interesting ones are the detection of planets around stars by measuring the small periodic variations in radial velocity they generate in their orbital motion, the radar measurement of speeds of vehicles, and the measurement of the speed of blood in arteries, which uses ultrasonic waves. Neither Doppler nor Fizeau could have imagined to what extent their discovery would have practical importance.

---

[65] One finds on the cover of the manuscript published in 1879 the following anonymous note: "It is important to reserve some room in the next issue of the Annals for the enclosed manuscript of Mr Fizeau, which is of considerable interest." (Académie des sciences-Institut de France, Fonds 64 J, dossier 8.10).
[66] *Secchi, A. (1870) *CRAS* 70, p. 903-6; see also *CRAS* 70 p. 1013, 1062-6 et 1213-4.
[67] See e.g. +Slipher, V.M. (1915) Spectrographic Observations of Nebulae, *Popular Astronomy* 23, p. 21-24.
[68] +Lemaître, G. (1927) *Annales de la Société des Sciences de Bruxelles* A,47, p. 49-59. Hubble, E. (1929) A relation between distance and radial velocity among extra-galactic nebulae, *Proceedings of the National Academy of Sciences of the USA*, 15, p. 168-173, accessible via http://www.pnas.org/content/15/3/168.full.pdf+html

Chapter 4
The velocity of light and electricity

Fizeau's apparatus for measuring the speed of light in 1849. The original device is lost, and Fizeau reused the telescopes for further experiments. Here we see a replica created by Paul Gustave Froment (1815-1865), probably commissioned by Jules Celestin Jamin (1818-1886) for his demonstrations at the Polytechnic School, where he was professor from 1852; it could possibly be the "big machine that was built under the direction of commissioners of the Academy [of Sciences], [which] will repeat the beautiful experiments of Fizeau with all the accuracy that commands the interest of science": this device is mentioned by Arago[69]. On the left, the drive mechanism of the toothed wheel that is seen in the foreground; a portion of the transmitting telescope to the rear; the telescope to reflect the light in the front. Compare with Figures 4.6 and 4.7. Lebée, Inventaire général/École polytechnique, Palaiseau.

---

[69] *Arago, F. (1854-1857) *Astronomie populaire*, t. 4 p. 425. We could not find any information on the fate of this apparatus.

4.1 A fertile period

Fizeau performed a considerable amount of scientific work during the early months of 1849. We are well informed about this activity through drafts and notes of experiments kept at the Museum of urban and social history of Suresnes, to which they were bequeathed.

At that time, Fizeau was obsessed to some extent with the possible applications of a very simple object: a toothed wheel rotating very rapidly. He was probably not the first to imagine that such an object could be used to measure the velocity of light. Moigno wrote that a priest named Laborde, a physics professor at the minor seminary in Corbigny (Nièvre), sent to Arago "there are five or six years," a letter that was lost "in the huge portfolio of Observatory"[70]. Moigno reproduced it in part and we see that it gave the principle of Fizeau's experiment that we will soon describe in detail: a light beam intercepted by a rapidly rotating wheel with holes arranged regularly on its periphery. The beam is thus chopped by this device. It is reflected on a remote mirror so that it comes back on itself. If the wheel turns slowly the beam has time to go back through the hole it crossed initially. If the speed is increased sufficiently, the reflected beam strikes the wheel on the interval between two holes, and it is stopped. If the wheel rotates even faster, the reflected beam passes through the next hole and becomes visible again, etc. This is the principle that Fizeau used, with the difference that his wheel had regular rectangular teeth on its periphery, which are easier to make than the holes of Father Laborde. Did Fizeau know about the idea? He said nothing about it. His student Cornu explained how by a strange path, which he reconstructed from conversations he had with Fizeau, the latter had the idea to use a toothed wheel[71]:

> "It is when thinking at Newton's theory of accesses[72] that he had the idea of this method; it assumes elongated molecules of light that turn like the spokes of a wheel, which explain in a somewhat childish, but very clear way, reflection, refraction and colored rings. They appeared to him as an achievable mechanical symbol. He pictured these long molecules rotating around their center, arriving sometimes by their tip to cross the surface of a new medium, sometimes

---

[70] °Moigno (1850), p. 1162-5. The letter has not been found since.
[71] *Cornu (1897) p. C13-14.
[72] See Newton, I. (1730) *Opticks*, 4th edition, London, Innis; 2d Book, part. III, Prop. XII, sqq. Accessible via http://www.gutenberg.org/ebooks/33504

by their side to reflect on it. Similarly, but conversely, a stick, rotating around its center, receiving a series of projectiles thrown along a straight line, could either intercept them, or let them pass, according to the phase of its movement. From there to the design of the method of the toothed wheel, it was only a step, and that step was quickly taken."

This is a fine example of the often unexpected way through which scientific ideas arise. The new method was only an extension of an idea that Galileo had before, and that Fizeau probably knew. Galileo stood at night at a considerable distance from another observer. Each had a lighted lantern he could cover. Galileo uncovered his lantern, then the distant observer uncovered his own one when seeing the signal. When Galileo saw the distant observer uncovering his lantern, he estimated the time elapsed between the time he uncovered his lantern and that when he perceived the light from his help. If the velocity of light was not so large, he could have measured it in this way. The use of a toothed wheel has obvious advantages with respect to this simple experiment, but the principle is the same.

In January and February 1849, Fizeau considered several applications of the toothed wheel at once: the measurement of the velocity of light, that of the velocity of electricity in a conductor and the comparison of the velocity of light in air and water. On January 20th, he filed a sealed envelope containing the principle of his measurement of the velocity of light to the Academy of Sciences[73]. He worked so fast at the design and implementation of these experiments that on the 17th of July 1849 he could write to Arago, who was then the all-powerful representative of physics in France thanks to its status as Permanent Secretary of the Academy of science, the following letter, the draft of which is preserved in Suresnes:

"Sir,
Held at home by indisposition, please excuse me if I do not present you myself the attached note relative to a new experiment that I have done in the recent days, in which I could see and measure the velocity of propagation of light between two relatively close stations, in Suresnes and Montmartre. As soon as I will be able to come to the Observatory, I hope to have the honor to speak to you about this and to supply for the want of clarity that the note might have,

---

[73] For a reproduction of the drawing inserted in this envelope, see Frercks (2000) p. 254.

because I have endeavored to do without figures so that it could be included in the Comptes rendus.

I work at the same time, with the collaboration of an inspector of telegraphs, Mr. Gounelle, to apply the same method to measure the speed of electricity in conductors, and I hope to be able to solve this issue in a more accurate way, less prone to objections, that Mr. Weathstone's [sic, for Wheatstone].

The same principles, slightly modified, give a new way to solve the problem of the velocity of light in refractive media, the problem you have posed in order to decide between the two theories of light and which should be accessible to experiment using rotating mirrors. The principle to which I come is more complicated but would require less extraordinary rotational movements [he was thinking of using one or two toothed wheels]. I have not made, however, any attempt in this direction and I will only take care of this if I have your formal invitation. Please accept, Monsieur, the expression of my deep respect."

As the three experiments that Fizeau mentions were conducted almost simultaneously, Annex 1 gives a fairly detailed chronology. We will discuss the first two in this chapter.

4.2 The first direct measurement of the velocity of light

While the principle of the measurement is simple, its implementation is not. The light has to be chopped by the peripheral teeth of a rapidly rotating wheel with variable speed, and the return beam to be observed. Fizeau quickly reached the design shown in the diagram of Figure 4.1. Oddly enough, he thought for a while to use two gears rotating in opposite directions, as we can see in several of his drawings, but he did not explain why or how the device would have worked, which is far from obvious.

The biggest challenge was to return the light beam that passed through the wheel exactly on itself. Today, we have a very simple device to do this, which considerably facilitates the reproduction of Fizeau's experiment: the cube corner, a set of three flat mirrors arranged in three perpendicular planes. This device did not exist at the time of Fizeau, who groped about placing a concave mirror behind the focus of a telescope, or what amounts to the same, a set lens/plane mirror or just a plane mirror (Fig. 4.2).

The velocity of light and electricity

FIGURE 4.1 – Autograph pen drawings of the emitting device of the experiment for measuring the velocity of light. The light of a lamp or that of the sun comes from the bottom, is sent to the axis of the telescope by a semi-transparent plate inclined by 45°, and is chopped by the toothed wheel located at the focus of the lens. The return beam is seen by the observer through the semi-transparent plate. Musée de Suresnes.

FIGURE 4.2 – Two autograph drawings in pencil showing how the light was reflected back at its arrival by the telescope on the right. Fizeau visibly hesitated on this critical issue. It is the "first way" (top), in which a lens whose focal point was the focus of the objective of the telescope and was followed by a plane mirror, which was finally retained with a simplification: the plane mirror was placed directly at the focus of the telescope (Fig. 4.3). A reticle could be substituted to it for alignment. Musée de Suresnes.

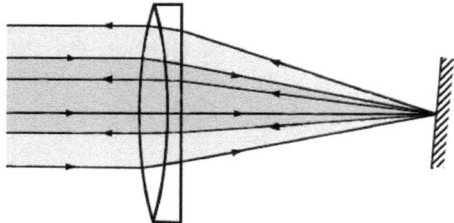

FIGURE 4.3 – The final solution used by Fizeau to reflect the light was a simple plane mirror placed exactly at the focus of the achromatic lens of the telescope. This diagram shows that the incident beam is reflected in the arrival direction even if the mirror is slightly tilted: the reflected beam is simply moved laterally without changing direction. Author's diagram.

On the 9th of February 1849, Fizeau tried the principle successfully, with a distance of 34 meters between the telescope containing the wheel and the one that reflected the light. But this distance was insufficient to obtain a scientific result. Fizeau, who lived at 5 rue Palatine in Paris, an inconvenient place, put the transmitting telescope in the gazebo of his parents' house in Suresnes (Fig. 4.4), and the reflecting one in Montmartre, in a place that we do not know but which is said to have been the window of the optical telegraph[74]. The distance between them, 8,633 meters, was estimated from a detailed plan for the region consulted with Auguste Félix Bruzard (1796-1855), the chief architect of the Paris Prefecture. Fizeau made many calculations to estimate the luminosity of the experiment and tried various light sources: oil lamp, ether lamp, Drummond light (which consisted of a piece of chalk heated white hot by a welding torch, see Fig. 1.4), and sunlight reflected by a heliostat, which might be the one that is kept in the Suresnes Museum (see the frontispiece of chapter 2). He used his calculations and his measurements to design a stellar photometer, which was however never constructed[75].

---

[74] From °Anonymous (1849) *Revue scientifique et industrielle* 36, p. 393-397, p. 394. Was it the Chappe optical telegraph, which was installed on the spire of the Saint-Pierre de Montmartre church? The distance between the Suresnes town hall and this place is longer than indicated by Fizeau by about 100 m, but this is probably within the errors.

[75] This photometer would have used, too, a rotating disk, but it would have had very deep triangular teeth. The reference light would have been decreased by bringing the beam to an appropriate distance from the axis of rotation of the disk, such that its intensity would have matched that of the beam received from the star.

The velocity of light and electricity

FIGURE 4.4 – The house of Fizeau's parents in Suresnes, in the 1880s. Much of the house, including in particular the gazebo where Fizeau had set his equipment, had been demolished, and the rest had become the Suresnes town hall from 1855 to 1886, which itself was destroyed in 1889. Musée de Suresnes, inv. 997.00.364.

— Only the Drummond's light at night and the sun during the day were intense enough for the success of the experiment. On April 12th, Fizeau finally saw the light beam from Suresnes reflected by the telescope he placed in Montmartre, and after some improvements, on April 25th, he said: "I [could] see with the naked eye a very bright spot amongst the lights of Montmartre. I think we will see it with a magnifying glass [i.e. with the eyepiece of the telescope]; after a few attempts made by illuminating the field I make sure that the experience will succeed with the disk."

— It is likely that he had already ordered the complete device from Paul Gustave Froment, one of the best manufacturers of instruments of the period (Fig. 4.5), because he wrote on May 18th that it would have been completely finished that evening. It cost 450 francs to Fizeau. Figure 4.6 is a diagram of the set-up, and an engraving in Arago's *Astronomie populaire* gives a detailed representation (Fig. 4.7). A description and a diagram are also given by Jamin[76].

---

[76] *Jamin (1885) p. 292-4 and pl. III.

FIGURE 4.5 – Paul Gustave Froment (1815-1865). Author's collection.

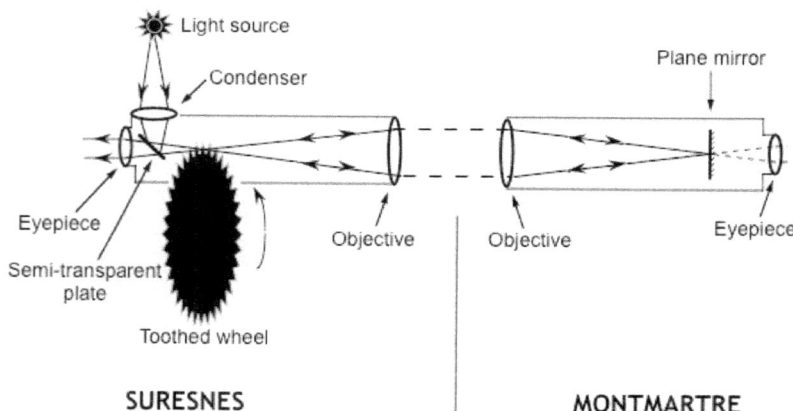

FIGURE 4.6 – Schematic diagram of Fizeau's apparatus. He used existing telescopes he modified. To the right, the telescope contained a plane mirror at its focus, which had the property of reflecting light in the direction of arrival. To place it exactly at the focus, Fizeau observed in a non-silvered part of the mirror the image of distant objects and moved it until the focusing was perfect. The alignment of the two telescopes was done by substituting a reticule with two crossed threads to the toothed wheel on the one hand, and to the mirror on the other. Diagram by the author

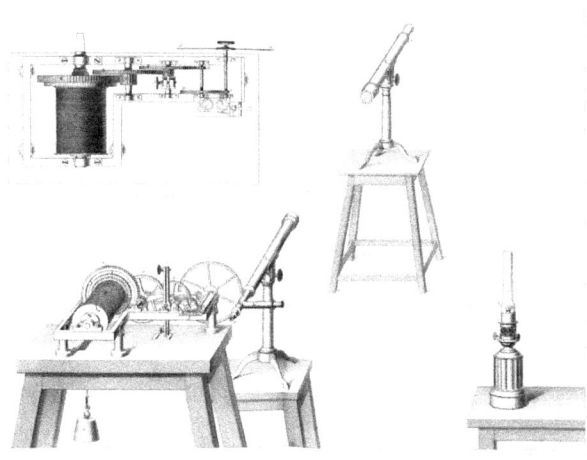

FIGURE 4.7 – The Fizeau's apparatus for measuring the velocity of light. The light beam from the lamp (in reality Drummond's light or that of the Sun reflected by a heliostat) was chopped by the toothed wheel. This wheel was driven by a clockwork mechanism with a weight, the top view of which is shown at the top left. The telescope sent the Suresnes beam to Montmartre, where another telescope with a mirror returned it to the first one. The returned beam crossed the toothed wheel and was examined through the eyepiece. A tachometer and a stopwatch made it possible to know the speed of the wheel. From Arago's *Astronomie populaire*, author's collection.

The first experiment took place on June 26th and 30th and July 3rd, using a cardboard disc with 997 teeth. Fizeau found a velocity of light of 331,400 kilometers per second from his measurements, but the results were quite dispersed and he decided to start over with a copper disc with 720 teeth, which must have been manufactured by Froment on a machine to divide circles. On July 8th and 9th, an occultation occurred when the disk rotated at about 12.6 revolutions per second, and the light returned fully for twice this rotation speed, for which the light passed back into the gap between the tooth following the one where it passed at emission; then it disappeared again for a triple speed (Fig. 4.8). Fizeau deduced from his measurements a velocity of light of 315,300 km/s (Fig. 4.9). He knew that this value was not very accurate, but he obtained a remarkable result: the first evidence from an experiment of the finite velocity of light, for which he found a value consistent with astronomical observations.

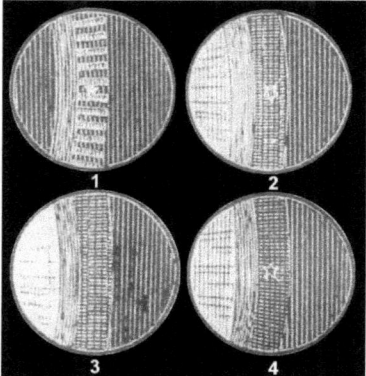

FIGURE 4.8 – What Fizeau was seeing in the eyepiece of his telescope; the light reflected from Montmartre is symbolized by a star. 1, the toothed wheel was motionless and the return beam passed between two teeth, the light appearing fully; 2, the wheel turned at low speed, the teeth could not be seen separately and the return beam was still visible, though attenuated; 3, the speed was such that the return beam stroke, the gap, between two teeth and was no longer visible; 4, the speed was even greater and the return beam became visible again. From Arago's *Astronomie populaire*, author's collection.

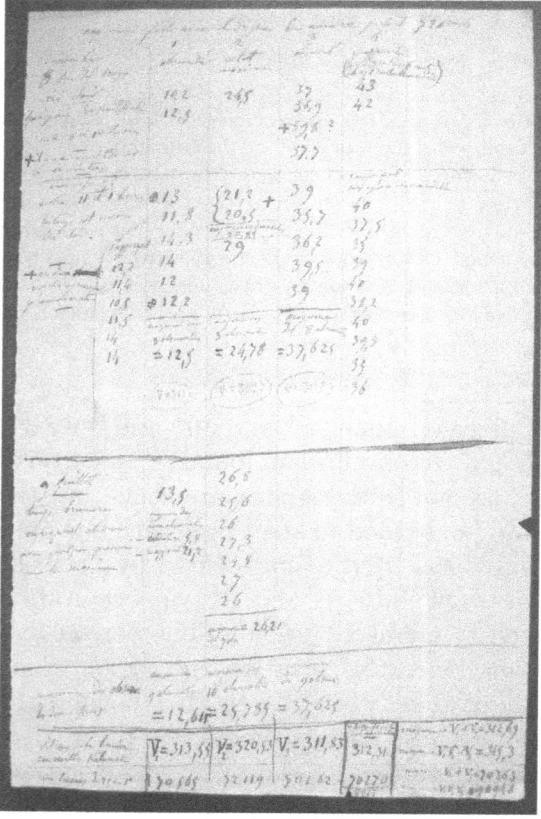

FIGURE 4.9 – Autograph manuscript showing the results of the experiment for measuring the velocity of light on July 8th and 9th 1849. These were night observations. Fizeau measured the number of revolutions of the wheel per second when the beam disappeared (left column), when it was fully visible again (middle column), and when it finally disappeared then reappeared again for a double speed (columns on the right). This second reappearance was hardly observable. The final result is given at the lower right: the velocity of light, 315.3 thousand kilometers per second, is the average of all the measurements. Musée de Suresnes.

The experiment and its result are described quite succinctly in a Note to the *Comptes-rendus de l'Académie des sciences*[77]: it is the one that accompanied the letter to Arago dated July 17th, which we quoted above. One reads there that the telescopes had a 6-cm aperture. To obtain all the details of the experiments, we have to read Arago[78], and even better Verdet[79].

4.3 The measurements of the velocity of light after Fizeau

The Fizeau's experiment immediately had a great impact in the scientific community and even in the general public. It received the honors of newspapers, including the *Journal des débats* where the usual scientific columnist, who was no other than Foucault, published a remarkably well written article on December 20th [80]. Foucault recalled that the velocity of light was already fairly well known by astronomical observations, but that this was its first direct measurement. He concluded:

> *The work of Mr. Fizeau, communicated in its simplest form, has nevertheless been welcomed with great favor; it is only fair; and the red ribbon that appeared recently in his buttonhole [the Légion d'honneur] demonstrates that the scientific spirit has survived the vicissitudes of our time. It is wonderful to hear Mr. Fizeau expose himself the advantages of his method and tell us that he was fortunate in his research. He has sometimes good luck, I admit, but it is the same luck that provides the rhymes to the poet and the discoveries to men of genius.*

Fizeau had therefore obtained the *Légion d'honneur* for his measurement, as well as Froment. He also received a three-year award from the five academies of the Institute in 1856 to reward "the discovery that honors ... or best serves the country", a very important award because its value was 30,000 francs[81]. Nevertheless, he remained modest and admitted that his measurement was not very accurate. We mentioned a

---

[77] *Fizeau (1849) *CRAS* 29, p. 90-93 et 132.
[78] *Arago (1854-1857) *Astronomie populaire*, t. 4, p. 417-425.
[79] *Verdet (1872) p. 658-663.
[80] *Foucault, L. (1849) in *Journal des débats* of 20th December, p. 1-2.
[81] *Delaunay, C.E. (s.d., 1864) *Annuaire du Bureau des longitudes pour 1865*, Paris, Gauthier-Villars.

device that was supposed to have been more efficient, which was built at the request of the Academy of Sciences, but we do not know if it actually served and what was its fate. It was down to the successors of Fizeau to improve his measurement.

In 1862, we find Foucault doing a new measurement, this time with a spinning mirror[82]. This was at the request of Urbain Le Verrier (1811-1877), the irascible director of the Paris Observatory. Le Verrier considered that if it was possible to accurately measure the velocity of light, we could, by combining it with the relatively well-known time taken by light to cross the Earth's orbit, get a better value for the dimensions of that orbit. On a light path of only 40 m, Foucault found a velocity of 298,000 km/s, which is much closer to the current value (299,792.458 km/s in vacuum) than that of Fizeau.

However, Cornu, who was the preferred student of Fizeau and tended to underestimate Foucault because of the quarrel between the two men (see the next chapter), believed that the spinning mirror method was not as good as that of the toothed wheel. He performed a new measurement with a toothed wheel and new equipment, which is preserved at the Paris Observatory: in 1872 he took measurements over distances of 2.5 km then 10 km through Paris, and found 298 500 km/s. This value was so close to that of Foucault that his reservation fell. In 1878, a new measurement between the Observatory and the tower of Montlhéry gave him 300,400 km/s, with an uncertainty that he estimated as 300 km/s. This was hardly better than Foucault.

This was the end of the French exclusivity in measuring the speed of light, although some attempts have been made after Cornu[83]. Albert A. Michelson and Simon Newcomb took over in the USA from 1878 to 1883, and obtained a value of 299,860 km/s (as calculated in vacuum) with a spinning mirror, with an accuracy they estimated at 30 km/s[84]. This seemed inconsistent with Cornu's measurement. It is in this context that the director of the observatory of Nice, Henri Perrotin (1845-1904), decided in 1897 to reinstall the device of Cornu, with the help of the latter[85]. The measurement took place in 1898, the distant station

---

[82] See Tobin (2003) chapter 13.
[83] *Wolf, C. (1885) *CRAS* 100, p. 303-9.
[84] Newcomb, S. (1883) *Publications of the United States Naval Observatory*, 1st series, 2, p. 107-230, 7 pl.
[85] See Bogaert, G. & Blanc, W. (2011) *Reflets de la Physique* n° 26, p. 20-22, accessible via http://www.refletsdelaphysique.fr/articles/refdp/abs/2011/04/contents/contents.html

being at La Gaude, about 15 km from the Nice observatory. It used much more powerful telescopes than Fizeau and Cornu in Paris. The value they found was 299,900 ± 80 km/s, in good agreement with the US one, but raising the dissatisfaction of Cornu because it did not quite agree with his previous findings: for this reason, it was only published in 1900[86]. This result caused some turmoil, so that new measurements were undertaken between Nice and Mont Vinaigre, the highest peak of Esterel, located 46 km away. The new value was 299,880 ± 50 km/s, again in good agreement with Michelson and Newcomb. It was published in November 1902, after the death of Cornu, so that there was no further controversy[87]. Perrotin was considering new measurements between Nice and Mount Mounier (2817 m) in the Mercantour, and also between Nice and a summit of Corsica, but due to his untimely death on the 29th of February 1904, they were never performed.

Thus the two methods to measure the speed of light, when properly conducted, can give results of comparable accuracy. In a final effort, Michelson combined them for even more precise measurements[88]. Figure 4.10 is a diagram of his apparatus, with which he obtained the velocity of light in a vacuum of 299,796 ± 4 km/s, in excellent agreement with the present value of 299,792.458 km/s. Michelson then programmed a measurement in vacuum in 1930, but he could not carry it out himself because he died the following year; it gave 299,774 ± 11 km/s. This was the last measurement using mechanical processes. Several laboratories, including one at the Observatory of Paris, succeeded in 1970 to simultaneously measure the wavelength and the frequency of a light source, in this case a laser in the infrared. The velocity of light being simply the product of these two quantities, it could be obtained with extraordinary precision, reaching 0.2 m/s. As time can also be measured with extremely high accuracy, the 17th General Conference on Weights and Measures decided to set the velocity of light in 1983 to 299,792,458 m/s exactly, and to derive the definition of the meter from it: the meter is no longer a fundamental unit[89].

---

[86] *Perrotin, H. (1900) *CRAS* 131, p. 731-734.
[87] *Perrotin, H. (1902) *CRAS* 135, p. 881-884.
[88] +Michelson A.A. (1924) Preliminary Experiments on the Velocity of Light, *Astrophysical Journal* 60, p. 256-61; (1927) Measurement of the Velocity of Light Between Mount Wilson and Mount San Antonio, *Astrophysical Journal* 65, p. 1-13.
[89] The official definition of the meter is: "The metre is the length of the path travelled by light in vacuum during a time interval of 1/299 792 458 of a second."

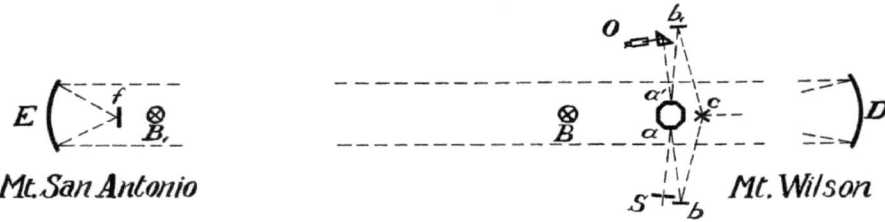

FIGURE 4.10 – Michelson's apparatus for measuring the velocity of light in 1926. The light of an electric arc entering the slit S, was reflected on the face a of a rotating octagonal mirror, which sent it to b then to c, where a mirror was placed at the focus of the reflecting telescope D: this telescope sent a parallel beam to the remote station E. Then the reflected light returned to D, which concentrated it in c, where a small mirror sent it to b_1. It fell on the octagonal rotating mirror in a', where it was observed with the eyepiece 0. The rotation velocity of the octagonal mirror was adjusted by stroboscopy from a tuning fork, so that the observed image appeared still, the mirror rotating by 1/8 turn during the round trip of light. Other spinning mirrors with 12 and 16 sides have also been used by Michelson. B and B_1 were for geodetic marks for measuring the distance between the two mountains (35 km). According to +Michelson (1927) *Astrophysical Journal* 65, p. 1-13, with the permission of the American Astronomical Society.

4.4 The velocity of electricity

As early as the eighteenth century, there were attempts to measure the velocity of electricity. This term refers to the speed at which an electrical disturbance moves along a conductor, not to the speed of electrons in a conductor when the current passes; the latter speed is very low[90]. Among the first experiments, the most complete was probably the one that William Watson (1715-1787) made in England in 1747. He measured the discharge velocity of a Leyden jar along a 3.5 km long chain formed by observers interconnected by iron wires and provided with timers, the current being returned by the ground. He found that the shock was felt at the same time by all observers, and concluded that the velocity of electricity was too large to be measured in this way.

In the following century, the problem got attention anew because of the advent of the electric telegraph. It was important to measure the

---

[90] Fizeau writes in a note dated 14 September 1849 (Muséum national d'histoire naturelle, Ms 3603-102): "If electricity is nothing else than a motion of ether, it seems necessary to accept that it moves freely in metals, the matter itself being unmodified as it is in acidic liquids where the propagation is only possible through some decomposition."

velocity of electricity to see whether the telegraph could be used over long distances. This is what wrestled with Charles Wheatstone (1802-1875) in 1834, when he was already thinking about installing an electric telegraph in England, what he achieved in 1838 between London and Birmingham. For his measurement, Wheatstone sent an electrical pulse of high voltage, resulting from the discharge of a Leyden jar, in a 400-m long circuit with three cuts, one near each electrode of the jar and one at the middle of the circuit. These three cuts were aligned on the same insulating plate, so that the three sparks that occurred at each pulse were visible by a single observer. Due to the propagation time, the spark of the middle of the circuit occurred with a small delay with respect to the other two, which were simultaneous. This delay being obviously very short, Wheatstone imagined to observe the three sparks by reflection on a mirror of polished steel rotating at about 800 revolutions per second (Fig. 4.11). This mirror was driven by a pulley with a belt run by the wheel of an impeller. As the mirror rotated slightly between the time of the two sparks near the generator and that from the middle of the circuit, one expected to see a slight shift of the image of the latter relative to the other two. The pulses were synchronized with the rotation of the mirror thanks to a rotary switch, so that the successive sparks appeared always at the same position and seemed to make a single picture because of the persistence of luminous impressions. Wheatstone noted that the sparks were not instantaneous, causing some elongation of the images seen by reflection on the rotating mirror. He observed a small delay of the spark from the middle of the circuit compared to the other two, and concluded that the electricity propagates at a higher speed than light, but he gave no value. This was a little disappointing, but at least it was clear that the electric telegraph could operate over long distances, and one could go into its construction.

The paper of Wheatstone[91] is very interesting, because it describes not only the invention of the method of the spinning mirror, but also its application to observing fast phenomena such as the vibrations of sound, which he had himself observed as well. We will see later that Arago imagined that it could be used to measure the velocity of light. Wheatstone also considered building a polygonal spinning mirror,

---

[91] Wheatstone, Ch. (1834) An Account of Some Experiments to Measure the Velocity of Electricity and the Duration of Electric Light, *Philosophical Transactions of the Royal Society of London* 124, p. 583-591, accessible via http://www.jstor.org/stable/108080

something that Michelson perhaps remembered nearly a century later. Finally, he had the intuition of the stroboscope, implemented for the first time, at least in France, by Foucault to control the speed of the spinning mirror in 1862.

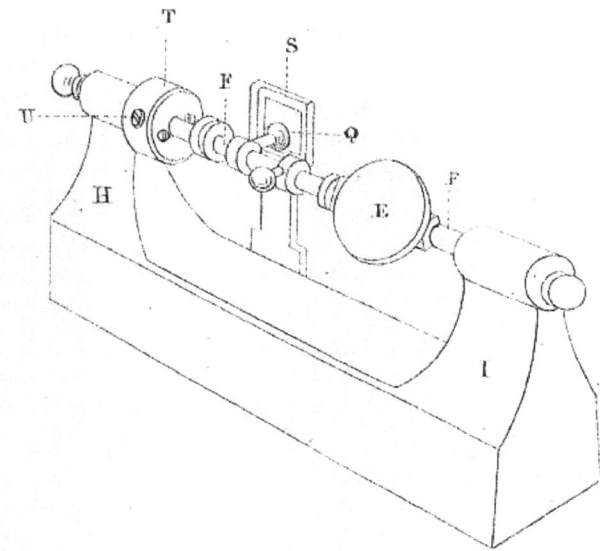

FIGURE 4.11 – Wheatstone's spinning mirror. E is the mirror of polished steel. It rotated around the axis FF thanks to the small pulley to the right of the cylinder T. This cylinder was a hollow box in which one could blow a lateral air stream: when the hole U passed in front of the jet one heard a sound whose frequency gave the speed of rotation of the mirror. However, this sound being too faint, the device was replaced by a cardboard strip (not shown here) attached to the shaft, which struck an obstacle at every turn (see a possible offspring in Fizeau, Fig. 3.3). The small ball Q passed at each turn near a fixed ball (not shown) fed by the high voltage generator; the corresponding spark, channeled by a thin horizontal slot through the mica plate S, opened the electric circuit at each turn, the axis of rotation being connected to the rest of the circuit by a permanent contact (not shown). Bibliothèque de l'Observatoire de Paris.

Fizeau was very familiar with Wheatstone's experiment, which Arago had certainly mentioned to him: Arago was in fact in Edinburgh, soon after this experiment, to attend a meeting of the British Association for the Advancement of Science, and was very impressed by the method of spinning mirror and its possibilities. He had considered the possibility of applying it to a decisive test of the nature of light, which we discuss in the next chapter. But Fizeau, who criticized Wheatstone's work, started the same study but using his toothed wheel rather than a spinning mirror. Notes dated 1st of March 1849, preserved in the Museum of Suresnes, show that he considered three different experiments:

— A measurement of the duration of an electric spark, the light passing through the teeth of a rapidly rotating disk;

— A measurement of the velocity of electricity using a rotating solid disk, with sparks making marks on the surface;

— Finally, the same measurement but using the toothed wheel whose teeth would make contacts.

Only the last principle came to realization. A note dated the 2nd of May 1849 describes it in more detail and contains the diagram whose draft is reproduced in Figure 4.12. To achieve this, Fizeau associated with Eugène Gounelle (1821-1864), who with Breguet had built the first electric telegraph lines in France. After some tests, the experiments used the telegraphic lines from Paris to Meulan, 47 km away, and between Calais and Lille, at a distance of 106 km. Between August and October 1849, the really serious experiments used the telegraphic lines from Paris to Rouen and from Paris to Amiens; the lines had two wires because there was no return by the ground; these wires were connected together at Rouen or Amiens. This provided a single loop, 288 km long in total for the Rouen line, partly made of iron and partly of copper, and an iron loop of 314 km for the Amiens line[92]. Several set-ups were used; the best one (Fig. 4.13) is slightly different from the initial arrangement shown in Figure 4.12. Fizeau and Gounelle found a velocity of electricity of about 180,000 km/s on the Rouen line and 100,000 km/s on the Amiens one, which is more reasonable than the result of Wheatstone. They attributed the difference of the velocities to the nature of the wires.

Fizeau criticized the experiments of the Americans Sears C. Walker (1805-1853) and Ormsby M. Mitchel (1809-1862), who found values of the velocity that were much smaller than his own[93]. But finally everyone was probably right: Michael Faraday (1791-1867) did measurements with all kinds of conductors, especially submarine cables, and found the most varied values, being unable to explain why[94]. It is true that if he discovered electromagnetic induction, he did not consider

---

[92] *Fizeau, H. & Gounelle, E. (1850) *CRAS* 30, p. 437-440.
[93] *Fizeau, H. (1851) CRAS 32, p. 47-48.
[94] *Faraday, M. (1854) *Annales de Chimie et de Physique* 41, p. 123-128.

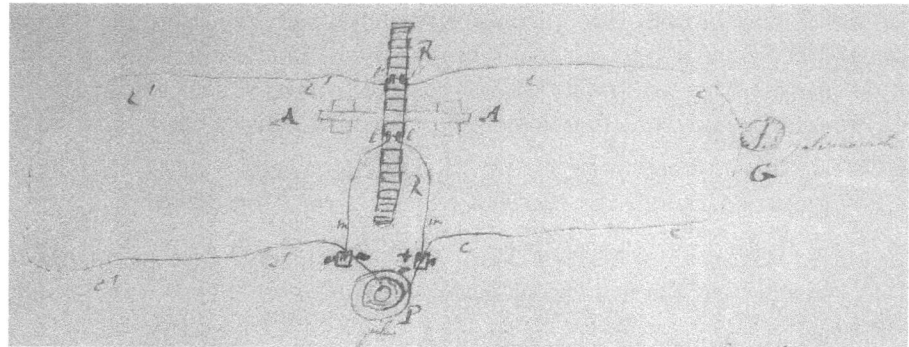

FIGURE 4.12 – Autograph ink drawing by Fizeau, dated May 2, 1849, illustrating the principle of the measurement of the velocity of electricity. Here are the explanations that accompany this scheme: "RR is a metallic toothed wheel, whose empty spaces are filled with a non-metallic material (stucco, ivory, wood); the wheel rotates about the axis AA and can make a large number of turns in a second (50, 100); l l and l' l' are small metal tongues forming a spring and intended to close the circuit when they touch the teeth. When l l are touching the metal the circuit is P l l P [a short-circuit]; when l l leave the metal the current is sent through the conductor c c G c c, and passes through the system l' l' [then the circuit c' c' c' c'] if these tongues are touching the metal; the two tongue systems can be adjusted so that the total circuit is complete [closed] during all the time that l l do not touch the metal. If the conductors are quite long, and the speed of the wheel such that the velocity of electricity is observable, the total circuit will not be complete [closed] during the same time. If the electricity arrives in l' l' after a time equal to the passage of a tooth, nothing will happen, the current will be zero." Musée de Suresnes.

self-induction, which, with the capacity and resistance of the cable, actually determines the velocity of propagation of an electrical disturbance. It was only much later, in the late nineteenth century, that this was understood and that it was possible to perform the appropriate calculations.

Fizeau undertook new measurements of the velocity of electricity, the outcome of which is not known by us. In doing so, in 1853 he discovered a way to avoid break extra-currents arising from the interruption of a circuit in an induction machine, which produce destructive sparks: it suffices to insert a capacitor between the two ends of this circuit, which temporarily stores the electrical energy from the self-induction of the machine[95]. This simple and ingenious invention is still in use today!

---

[95] *Fizeau, H. (1853) *CRAS* 36, p. 418-421.

The velocity of light and electricity 63

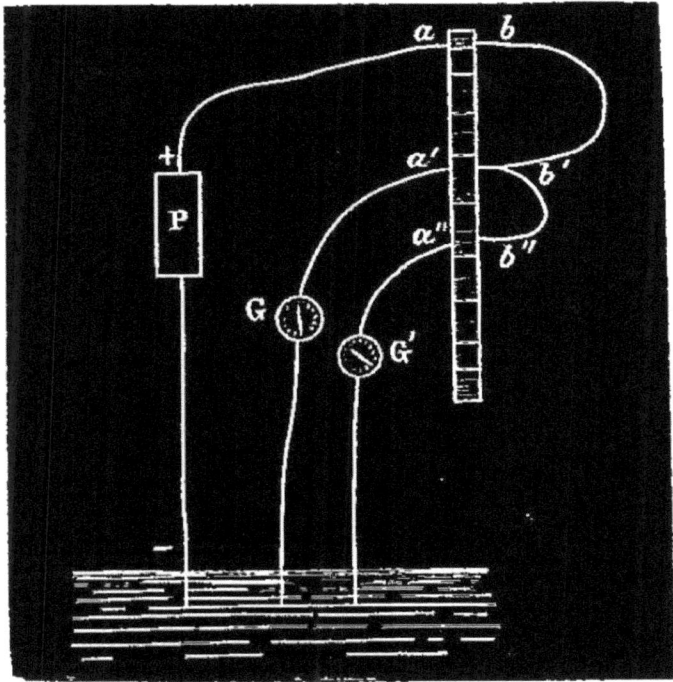

FIGURE 4.13 – The final set-up of Fizeau and Gounelle to measure the velocity of electricity. The wheel, 50 mm in diameter, had 36 divisions of boxwood, thus insulating, and 36 of platinum. It was driven by a device due to Froment, equipped with a meter for measuring the rotational speed. There were now three sets of two tongues forming contact: $a\,b$, $a'\,b'$ and $a''\,b''$. The telegraph circuit (in reality very long) was between the tongues b and b', the tongues b' and b'' being interconnected. The arrangement was such that when the tongues $a\,b$ were in contact with a conductive tooth, it was the same for $a'\,b'$ whilst $a''\,b''$ were on an insulating interval. The negative pole of the battery P is grounded, the other communicates with the telegraph circuit via $a\,b$: the current is chopped by the rotation of the wheel. At rest or with a low rotation speed of the wheel, the current flowing in $a\,b$ also goes through $a'\,b'$, operating the galvanometer G, while it does not go through $a''\,b''$, the galvanometer G' indicating no current. For a certain, greater, speed of rotation, depending on the propagation time in the telegraphic circuit, the current no longer passed through $a'\,b'$ but through $a''\,b''$ instead, G being at rest and G' actuated as on the figure. It was the reverse for a double rotation speed; G' was activated again for a triple speed, but the signal was then fainter. From Verdet (1872), Bibliothèque de l'Observatoire de Paris.

Chapter 5
The "crucial experiment":
the velocity of light in air and water

Breguet's spinning mirror, built in 1844 for the "crucial experiment" of Arago. Bibliothèque de l'Observatoire de Paris.

5.1 The project of Arago

On December 3rd 1838, Arago presented to the Academy of Sciences the project, of an experiment that, according to him, should have decided the nature of light definitively[96]:

"I propose to show in this paper, how it is possible to determine unequivocally whether light consists of small particles from the radiating body, as Newton wanted and as was admitted by most modern scientists; or whether it is simply the result of waves in a very rare and very elastic medium that physicists have agreed to call the Æther[97]."

Indeed, the controversy between the two theories of light at this time was still alive, although the wave theory imagined in the seventeenth century by Christiaan Huygens (1629-1695) and greatly developed in the early nineteenth century by Thomas Young (1773-1829) and by Augustin Fresnel (1788-1827) with the support of Arago, clearly was taking over despite the reputation of Newton. The latter had imagined that light was composed of small particles with mass, and explained the refraction, when light passes through a medium with a higher refraction index, by an attraction of these particles by this medium (Fig. 5.1).

Arago proposed to determine whether the velocity of light is greater in water than in air, which would verify the theory of Newton, or whether it decreases when the light beam enters the water. For this, he planned to use the Wheatstone's method of the spinning mirror, which we described in the previous chapter. Two parallel beams coming from an electrical spark would be sent on a rotating mirror; one beam would propagate in the air while the other would go through a tube filled with water. Then they would arrive on the rotating mirror, which would reflect them towards the observer with a telescope. One of the beams, being propagated more slowly than the other, would meet this mirror later, which would have had time to turn slightly, so that the observer would see the corresponding reflected beam deflected more than the other. Arago made small calculations which took into account the absorption of the light in water, which limited the length

---

[96] *Arago, F. (1838) *CRAS* 7, p. 954-65.
[97] It was inconceivable at the time that a light wave could propagate without some material support, hence the introduction of æther. Newton did not need it, but he nevertheless postulated its existence.

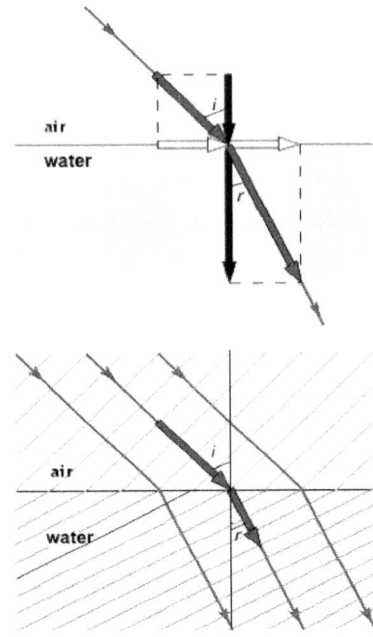

FIGURE 5.1 – Refraction. Top: to explain refraction, the Newtonian emission theory assumed that the velocity of light increased perpendicular to the surface of separation when entering into the dense medium (e.g. water), due to the attraction of particles of light by this medium, while the component of velocity parallel to this surface did not change; the velocity of light was thus larger in the dense medium. In this case, the relationship between the incidence angle i and the refractive angle r was written $\sin i / \sin r = n = v_2 / v_1$, where n was the refractive index, and v_1 and v_2 respectively the velocity of light in air and in water. Below: the wave theory, which is that which we know today, implies the continuity of the wave planes perpendicular to the direction of propagation of the light, when going from one medium to another. We see that in this case the velocity of light is smaller in the denser medium, where the wave planes are tighter. then we have $\sin i / \sin r = n = v_1 / v_2$. Author's drawing.

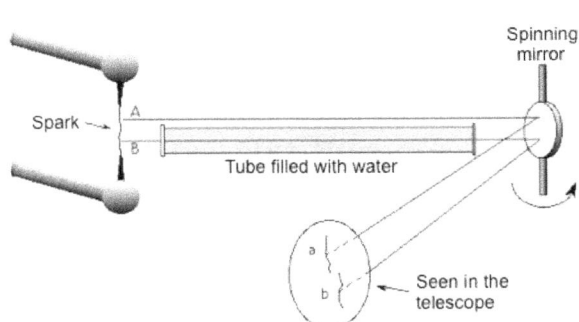

FIGURE 5.2 – Arago's project to compare the velocity of light in air and in water. Explanations in the text. Author's drawing.

of the water tube to a dozen meters. He found that a mirror rotating at 1,000 revolutions per second would allow him to see clearly which beam was the slower. However, the spark may have occurred at any stage of the rotation of the mirror, so that the chances of the observer to observe the reflected light were very small.

To realize the project, Arago had several devices with spinning mirrors built by Antoine Louis Breguet (1776-1858). One of them contained no fewer than three mirrors rotating at the same speed: the beam reflected on the first mirror ended after another reflection on the second, then on the third, so the deviation was tripled for the same rotation speed. But the light was weakened by these successive reflections. Breguet also built the instrument preserved at the Paris Observatory (picture of the frontispiece of this chapter); the mirror could turn at 2,000 to 3,000 revolutions per second. The trials lasted for a long time, and eventually Arago, who was already strongly feeling the diabetes and whose sight was deteriorating, gave up the search. This is why in 1850 he published the principle of the measurement, to mark his priority and hoping that someone would take it over[98].

Arago apparently did not understand Wheatstone's trick of synchronizing the electric spark with the rotation of the mirror. If he had done it, the light would have been returned in the same direction by the spinning mirror and it would have sufficed to place the observer there to see the phenomenon. However it might have been difficult to achieve perfect synchronization. Anyway, to increase the chances of seeing the spark in the telescope, Arago had proposed to dispose several observers around the rotating mirror and to use 8-10 mirrors instead of one on the rotary support. He also imagined to use, instead of water, carbon disulfide CS_2, which has a very high refractive index.

5.2 Fizeau and Foucault take up Arago's experiment

Fizeau was a familiar of Arago, and it is not surprising that he intended to take up his experiment. At the end of his 1850 article quoted earlier, Arago made reference to the letter addressed to him by Fizeau on the 17th of July 1849, a letter that we reproduced in the previous chapter and by which he informed Arago that he has thought about this experiment. This letter contains the following sentence: "I have not made, however, any attempt in this direction and I will only take care of this if I have your formal invitation." Arago also wrote in his article: "This loyal reservation could only add to the esteem that

---

[98] *Arago, F. (1850) *CRAS* 30, p. 489-95. This article contains a detailed history of Arago's project.

the character and the work of Mr. Fizeau inspired me, and I hastened to authorize Mr. Breguet to lend him one or several of my rotating mirrors."

In reality, Fizeau had first considered using his toothed wheel for this experiment, as shown by a manuscript dated 11th of February 1849, kept in the Suresnes Museum (Fig. 5.3). However, he must have realized later that the distance between the two telescopes and the length of the tube containing water should be very large for obtaining a significant effect, and that the project was actually not feasible. Conversely, the spinning mirror method requires no large distance and Fizeau decided to use it by taking up Arago's attempts in a different form, which is shown in an undated sketch preserved in the Archives of the Academy of science (Fig. 5.4).

But a competitor arose, who was no other than Foucault. Foucault came to Arago to ask permission to realize the experiment: Arago wrote in his note to the Academy of Sciences the 29th of April 1850: "Foucault, whose inventiveness is known by the Academy, came himself to inform me of his desire to submit to the test of experiment the changes he wanted to make to my apparatus." Arago did not appreciate the intervention of Foucault, but what could he do? Finally he gave up and ended with the following sentence: "I can only, given the present state of my sight, accompany by my wishes the experimenters who want to follow my ideas."

FIGURE 5.3 – The first of Fizeau's ideas for comparing the velocity of light in air and in water. A hole i in a screen placed at the focus of the telescope (*lunette*) N° 1 is illuminated by a light source. The resulting parallel beam is sent to the telescope N° 2, which has a mirror at its focus. The reflected beam arrives at the point i' of the screen. The light coming from i' would be reflected to i, hence the name "paired images" given by Fizeau to i and i'. If the screen is now the edge of a toothed wheel, i and i' can be arranged to be in the space between two teeth. If the two telescopes are sufficiently distant from each other and if the wheel turns rapidly enough, the light from i comes late to i', and vice versa, so there might be an occultation. By placing a tube full of water on one of the optical paths, the delay would be different and one could see in which case it is the larger. Musée de Suresnes.

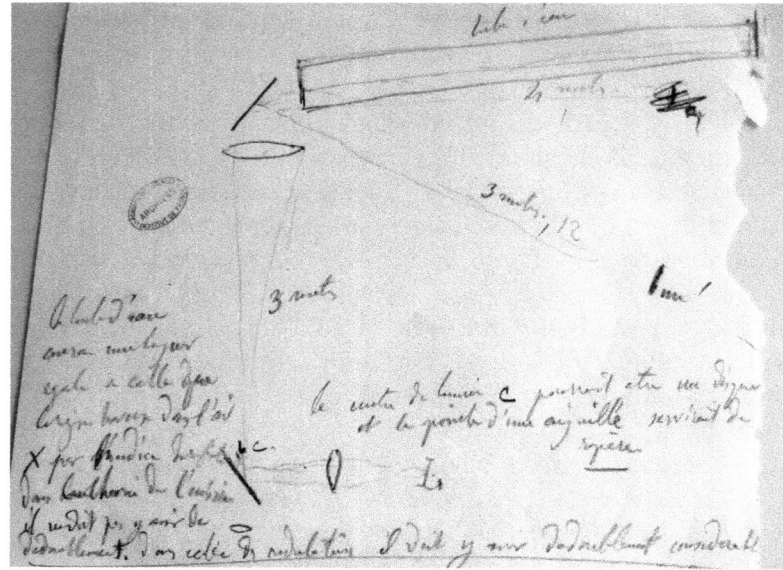

FIGURE 5.4 – Autograph scheme by Fizeau, undated but certainly later than that of Figure 5.3, showing an alternative set up for comparing the velocity of light in air and in water, close to the final realization. This time, a spinning mirror is used, schematically shown at the top left. A lens would produce the image of a point source of light L in c, after reflection on a 45° semi-transparent plate. Then the light would fall on the turning mirror and, depending on its position, would pass through air or water, be reflected by a flat mirror and returned to c, where it would be observed through the 45° semi-transparent plate. Fizeau observed the displacement of the image relative to a reference point (the point of a needle), when the mirror was rotated very quickly. The displacement was greater when the light had passed through the water than through air, hence a duplication of the image. Académie des sciences-Institut de France, Fonds 64.1, dossier 8.14.

In truth, our two experimenters had previously decided to collaborate. They were then faced with a big problem. The light from an electric spark was certainly insufficient for the experiment, and they had to give up the possibility of synchronizing the light source with the rotation of the mirror, if it was ever considered. The idea of Arago was, as we have seen, to have a circle of many observers around the spinning mirror in the hope that one of them would have the chance to see something. The German astronomer Friedrich Bessel (1784-1846), who was aware of Arago's project, had proposed to return the light that had been reflected by the spinning mirror through a plane mirror: a single observer would have then been sufficient.

Another advantage of Bessel's idea, that is sometimes wrongly attributed to Foucault or Fizeau, is the following: during the return of the beam, the mirror would have turned slightly and the beam reflected a second time by the rotating mirror would be slightly deviated: the measurement of this deviation would give the velocity of light. One recognizes here the principle that will invariably be used for the measurements of the velocity of light with a rotating mirror, an application of which we already see in the diagram of Figure 5.4.

However, we note that the plane mirrors of Figure 5.4 would have returned the light on the spinning mirror only during the very short time when the beam would have fallen perpendicularly to their surface, the light being lost at other times. Then Fizeau had a great idea[99]: he realized that if we replace Bessel's plane mirror by a concave mirror, the beam will return to the rotating mirror for the entire time it will sweep the concave mirror, and we gain a lot in terms of luminosity.

5.3 The steeple chase

It was then that Foucault, who now wanted to make the experiment alone, argued with Fizeau as evidenced by the letters of 22-24 April 1950 reproduced in Appendix 3. Fizeau was ulcerated and the two men parted; their quarrel was final. Foucault joined Gustave Froment to realize his camera, while Fizeau was working with Louis Breguet (1804-1883), whose father had built the rotating mirror, that he was going to use, for Arago. It is likely that these devices were ready before the quarrel. Under the terms of Cornu, "it was between the two rivals a real steeple chase to achieve the goal proposed by Arago; both arrived at the same meeting of the Academy (May 6, 1850) with nearly identical devices, demonstrating their fruitful past collaboration[100]; but, more fortunate than his rival, who had only succeeded in verifying the effectiveness of his apparatus, Foucault announced that his experiment had been successful: he had seen the deflection of the light ray passing through the water tube, and it was larger than the deflection through an equal length of air; the emission theory was therefore condemned, and the

---

[99] According to Cornu (1900) Sur la vitesse de la lumière, in *Recueil des travaux et discours d'A. Cornu* ... Vol. 5, accessible via http://jubilotheque.upmc.fr: see the note at the bottom of p. 6 to 8.
[100] *Foucault, L. (1850) *CRAS* 30, p. 551-60; *Fizeau, H. & Breguet, L. (1850) *CRAS* 30, p. 562-3.

wave theory demonstrated by a direct experiment, according to Arago's plan. Foucault thus undoubtedly came first to the purpose so desired."

The details of the competition are told by Tobin (2003). The delay of Fizeau and Breguet, who worked in the vast hall of the meridian of the Paris Observatory (today Cassini room, Fig. 5.5), was due to the fact that the water, contained in a zinc tube, was polluted and not transparent, so they were forced to build another tube made of crystal. They announced their success, which confirmed that of Foucault, on the 17th of June 1850, seven weeks after him[101]. Newton's theory had lived.

FIGURE 5.5 – The hall of the meridian of the Paris Observatory, now named Cassini room, where Fizeau and Breguet made their experiment. The meridian of Paris is materialized by a bronze line inserted in the middle of the room. The image of the Sun is projected on it at the local noon. The measurement of the velocity of light by Foucault in 1862 was also made in this room. Bibliothèque de l'Observatoire de Paris.

---

[101] *Fizeau, H. & Breguet, L. (1850) *CRAS* 30, p. 771-4.

The final experimental set up of Fizeau and Breguet (Fig. 5.6) was much like that of Foucault (reproduced in Tobin, 2003). The rotation speed of Breguet's mirror could reach 1,500 revolutions per second, thanks to helical gears, a novelty at the time. Fizeau and Foucault both used the sunlight reflected by a heliostat, which caused them a lot of trouble due to unfavorable weather.

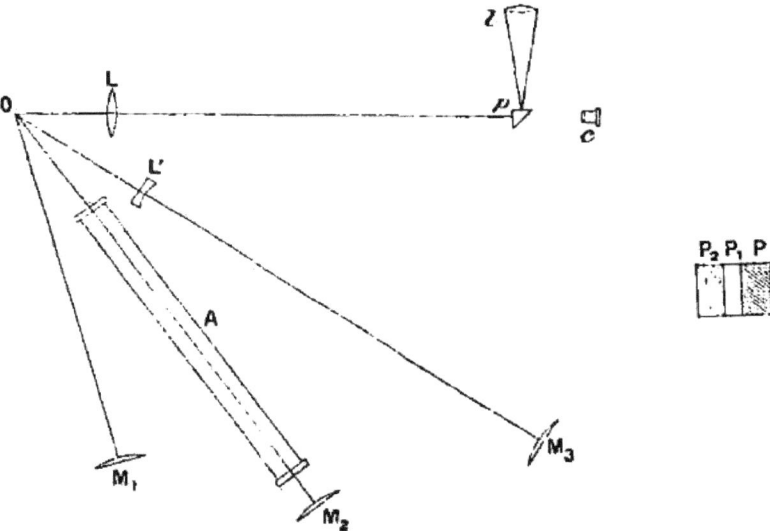

FIGURE 5.6 – Fizeau's set-up for comparing the velocity of light in air and in water (compare to Fig. 5.4). Sunlight illuminated the total reflection prism p, then fell onto the rotating mirror located at O. The latter sent the light to M_1, which was a plane-convex lens silvered on its rear plane face, equivalent to a concave mirror. It reflected the light onto the spinning mirror O. In another position of the rotating mirror, the light was sent into tube A filled with water, was reflected by M_2 and returned to the rotating mirror after a second crossing of this tube. If the mirror rotated slowly, the reflected light fell exactly on the prism p and Fizeau could only see its dark rear side P in the eyepiece c. If it ran faster, the light was deflected and that which had been reflected by M_1 fell partly outside the prism at P_1, and that reflected by M_2 at P_2. P_1 and P_2 were both seen in the eyepiece, as shown in the small figure to the right. Fizeau also used alternatively a reflection on a third reflector M_3 with an optical path length equal to that in the M_2, so that the two images were superimposed near the prism P regardless of the speed of the mirror: this obviously supposed that the result had been foreseen. From *Mascart (1893, Fig. 325 p. 90).

Chapter 6
The drag of æther

The protagonists of the story told in this chapter. The date of their work is indicated. The first one, John Michell (1784), is missing because there is no portrait of him known. Library of the Paris Observatory for Arago and Fresnel, author's collection for Fizeau, official photographs of the Nobel Prize (Wikimedia Commons) for Michelson, Einstein and von Laue.

This chapter requires careful reading, because the old notions of light and its propagation are very different from those prevailing today, and we are not always familiar with them.

6.1   Act 1: Michell, Arago and Fresnel

In 1784, the English scientist John Michell (1724-1793) wrote[102] that, in the Newtonian theory of light, which was universally accepted in England at that time, the particles of light emitted by a star could be slowed by the gravitational attraction of the star: the velocity of light would thus depend on the mass and radius of the star[103].

Michell also considered what happened when the light slowed in this way when it entered a dense transparent medium. We saw in the previous chapter that the explanation of refraction in Newton's corpuscular hypothesis implied that the velocity of light increased as it entered such an environment. Michell made the arbitrary, but logical assumption, that this increase was due to some attraction by the dense medium, which did not depend on the initial velocity of light but only on the nature of this medium. He then made a small calculation (Box 6.1) showing that the angle of refraction must then depend on the initial speed of light.

Box 6.1: Michell's calculation, transposed into modern notation.

Let v_1 be the velocity of the incoming light, and v_2 its velocity in the dense medium. The relation between the angle of incidence i on the separating surface and the angle of refraction r is written, in the Newtonian theory (see Fig. 5.1 and its caption):

$$\sin i \,/\, \sin r = v_2/v_1$$

The relation between v_2 and v_1 is written:

$$1/2 \; mv_2^2 = 1/2 \; mv_1^2 + A$$

---

[102] Michell, John (1784) On the Means of Discovering the Distance, Magnitude, &c. of the Fixed Stars, in Consequence of the Diminution of the Velocity of Their Light... *Philosophical Transactions of the Royal Society of London*, 74, p. 35-57. Accessible via http://www.jstor.org/stable/info/106576

[103] Michell stated that if the mass and dimensions of the star are quite large, the light cannot get out, which foreshadows the idea of black holes.

where m is the mass of the light particle and A a constant characteristic of the dense medium, which represents the increase of kinetic energy of the particle when it enters this medium. We can also write this equation:

$$v_2 = (v_1^2 + 2A/m)^{1/2}$$

This relation is non-linear, so that v_2 is not proportional to v_1 and the ratio v_2/v_1, hence the angle of refraction, varies if the velocity of the incoming particles changes.

We can thus, in this theory, see whether the velocity of light from different stars is not the same, by observing the corresponding change in refraction, for example by measuring the deflection of the light by a prism.

Michell does not seem to have put this idea to execution, but the young Arago, who heard about it, worked on it from 1805. He first used a prism with an angle of 45', whose dispersion was small enough to allow him to work in white light. He observed various terrestrial sources of light with and without prism (stars, the Sun, the Moon and planets) and found that the deviation of their light through the prism was always equal to 25', with variations by 5 seconds of degree at most that he considered rightly as due to measurement errors: he deduced that the velocity of light from all these objects was the same within 1/480. In 1810, he resumed these measurements with an achromatic prism giving a deflection of 10° (Fig. 6.1), then with two prisms in tandem giving a total deflection of 22° 25'. These prisms were placed in front of one half of the objective of a repeating circle, most likely that of Fortin, which is currently in the collections of the Paris Observatory (Fig. 6.2). Again, the deviation was the same for all stars!

Arago then had another idea: he observed the same star at different times of the year. Even if the light emitted by all stars had the same velocity, he hoped to see velocity variations in the light he received: he believed in fact that the orbital velocity of the Earth added or subtracted to the initial velocity of light. He hoped to detect these changes, but he still saw nothing[104].

---

[104] Although it was read before the Academy on 10th of December 1810, the manuscript of Arago, temporarily lost, was only published in 1853 "without changing a single word": *Arago, F. (1853) CRAS 36, p. 38-49.

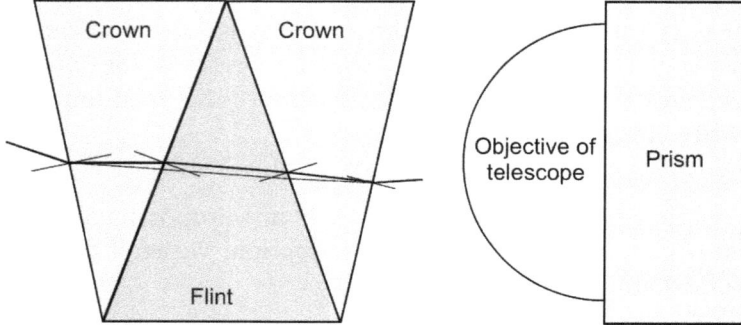

FIGURE 6.1 – Arago's achromatic prism. The present version (there were several) comprised three prisms. The path of two light rays is indicated by way of example. Lines perpendicular to the surfaces are drawn. The dispersions of flint and crown compensated, so that finally the light was deflected by the same angle regardless of its color, which allowed him to observe it in white light. Arago put an achromatic prism in front of one half of the objective of the telescope, as shown at the right, to measure the deviation by successively observing the star with and without the prism. Author's drawing.

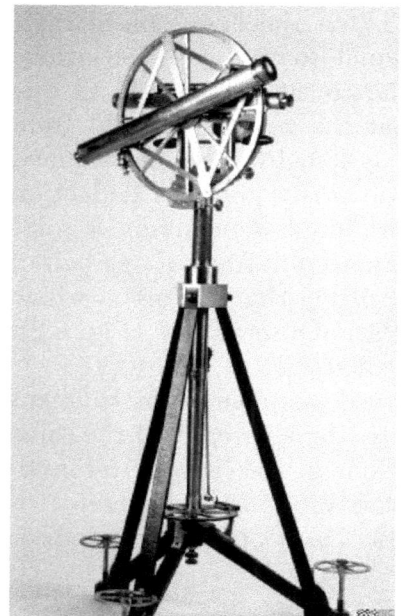

FIGURE 6.2 – Fortins's repeating circle, ca. 1806. Bibliothèque de l'Observatoire de Paris

This negative result set into turmoil the physics community, which was absolutely convinced of Newton's theory being true. To solve the problem, Pierre-Simon Laplace (1749-1827), followed by the young Arago, who he then influenced strongly, decided that the eye was sensitive only to light grains with a given velocity, an ad-hoc explanation to catch the club. Arago did not remain Newtonian for long. In 1818, he was working with Augustin Fresnel (1788-1827), who he had invited to the Observatory, and became increasingly convinced of the validity of the wave theory of light. Thinking back to his 1810 measures, he asked Fresnel if he could interpret their result in the context of this theory. We saw in the previous chapter that the propagation of light was then conceived "as the result of waves in a very rare and very elastic medium, which physicists have agreed to call the æther" (Arago in 1838). It was in this framework that Fresnel reasoned.

One problem is that Arago's prism did not allow him to measure the velocity of light if the wave theory was correct, so that one could not interpret the observed absence of deflection by the prism simply[105]. Fresnel found an interpretation nevertheless, that he published, as a letter to Arago, in the *Annales de Chimie et de Physique*, of which Arago was one of the two Editors in chief[106]. The demonstration is rather convoluted and the figure is incorrect, but other authors have confirmed the result. The box contains a recent analysis of Fresnel's reasoning[107]. The explanation assumed that instead of being dragged in the glass with its velocity u, which was that of the Earth, in which case this velocity would simply be added to that of light, the æther was dragged in the glass with the velocity $u(1 - 1/n^2)$, where n was the refractive index of the glass. In this case, the velocity of light in the glass of the prism would have been $v_{\text{glass}} = v/n \pm u(1 - 1/n^2)$, according to the direction of movement of the Earth with respect to that of the light (it is assumed here for simplicity that these motions are co-linear): the + sign corresponds to the case where the movement is in the same direction as the light.

---

[105] In the modern interpretation, whatever the motion of the Earth the light enters the prism with the invariable velocity c, which does not depend on u, and its velocity in the prism is always c/n. Strictly speaking, the light arrives with a slightly lower velocity since the prism is in the air; but this velocity still does not depend on the source and on the movement of the prism.
[106] °Fresnel, A. (1818) *Annales de Chimie et de Physique*, 9, p. 57-66; reproduced in °*Œuvres complètes de Fresnel*, publiées par H. de Sénarmont, E. Verdet et L. Fresnel, Paris, Imprimerie impériale, tome 2 (1868), p. 627-636.
[107] From *Costabel (1989).

Box 6.2. Fresnel's interpretation of the observation of Arago

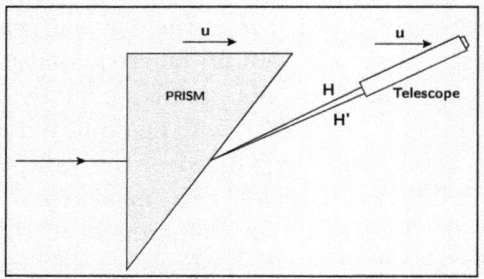

FIGURE 6.3 – Author's drawing, after Costabel.

Let us remember that in the nineteenth century, the light was supposed to propagate in an æther that filled the universe and was the absolute reference that Newton had proposed, with respect to which the Earth was moving. Let us reason in this framework. Suppose (Fig. 6.3) that the light comes from the left from a star from which the Earth, and thus Arago's prism, is moving away with velocity u. The light enters the prism perpendicularly to a face with the velocity v. Inside the prism, the light would propagate at velocity v/n, n being the refractive index of the glass, if this prism was stationary with respect to the æther; but this velocity might be a little different due to the motion of the prism (this is what Fresnel sought to know). At the exit, the beam is deflected by the prism in the direction H, and propagates again in the immobile æther. As the telescope used to observe the beam moves relative to this æther with the velocity u during the travel time of the light from the exit of the prism, we do not see the beam in exactly the direction of H, but in a neighboring direction, H': this is aberration (which does not depend on the distance from the prism to the telescope). Since Arago always observed the same deflection by the prism regardless of the velocity u, this means that the light does not propagate inside the prism with the velocity v/n, but with slightly different velocity v_{glass}, which modifies the deflection of light by the prism so as to compensate for aberration. Fresnel interpreted this by a "partial drag" of the æther inside the prism at a velocity slower than u. The calculation gives: $v_{\text{glass}} = v/n \pm u(1 - 1/n^2)$, according to the direction of movement of the Earth with respect to that of the light.

Upon receiving Fresnel's letter, Arago had already lost interest in the problem, at least temporarily, because he was engaged in work with Ampère about electromagnetism. Moreover, neither he nor Fresnel really seemed to believe in the explanation, which they never mentioned later: in particular, we would have expected that Arago would have said a few words about it in 1853 when publishing his 1810 article, but it is not so. Things therefore remained unchanged until the intervention of Fizeau.

6.2   Act 2: Fizeau and Michelson

In 1816-1818, Arago and Fresnel had done remarkable interferometric measurements of the refractive index of wet and dry air, with an apparatus (Fig. 6.5) whose principle is illustrated in Figure 6.4. The goal was to see whether humidity affected the refraction by the atmosphere, a major problem for the accurate measurement of the position of the stars[108]. The result was that the effect of humidity was very small in practice.

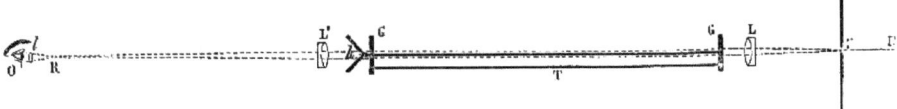

FIGURE 6.4 – Diagram of Arago's interference apparatus for measuring refractions. The source of light illuminated the slit on the right; the transmitted light formed a parallel beam after passing through the lens L. This beam was separated by slits into two parallel beams that the lens L' reunited in R, where interference fringes were observed with the eyepiece l (there was no need for Young slits in this device because the two beams were parallel, provided that the GG windows were of excellent optical quality). The upper beam passed through the external humid air, the lower one in a square tube T, 1 meter in length, where the air was dried. The oblique blades k could be rotated to compensate for the displacement of the fringes, and the measurement was done from their angle of rotation. Bibliothèque de l'Observatoire de Paris.

---

[108] See Lequeux (2016), chapter 10.

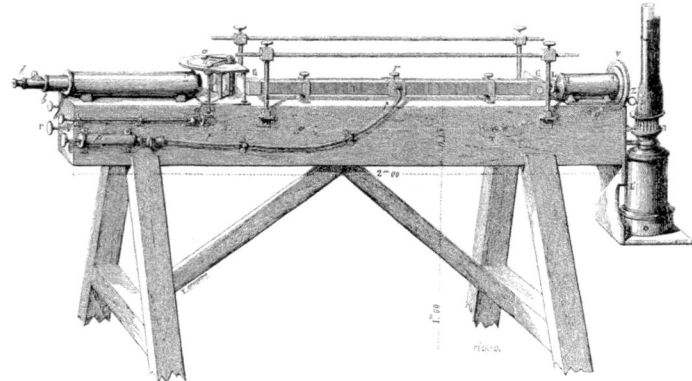

FIGURE 6.5 – Arago's interference apparatus for measuring refraction. Bibliothèque de l'Observatoire de Paris.

Another version of the device shown Figure 6.4, commissioned in 1816, had two identical tubes, one containing dry air and the other moist air. In 1852 Arago decided to resume the measurements with an increased precision and had a similar device built with two tubes 10 meters long for this purpose, that he installed in the meridian room of the Observatory[109]. But he was now almost blind, and he asked Fizeau to do the measurements, which he did "with the remarkable accuracy he brings to all these works".[110]

If Arago asked Fizeau to do the work, it is because he knew that he had used, the previous year, and perhaps even before, a very similar device to study the variation in the velocity of light in a moving medium. In fact, we must go back to an earlier time to know the genesis of this experiment, probably the most difficult and the most remarkable of all those done by Fizeau. Indeed, a sealed envelope deposited by Foucault at the Academy of Sciences on May 27, 1850 indicates that Fizeau and himself had tried to measure the drag of the æther by air in motion, but without success[111]. Foucault wrote in it:

"Due to the impossibility to see elsewhere than on the stars an aberration phenomenon due to the earth's translational motion in

---

[109] There is an image of this apparatus in Moigno's *Cosmos* (1854), reproduced in *Costabel (1984) *La vie des sciences* 1, 235-249, but it is of poor quality and we did not find it useful to reproduce it here.

[110] *Arago, F. (1854-1862), t. 11, p. 702-732, see p. 723-732.

[111] This document is reproduced by *Costabel (1984), who added interesting comments.

space[112], *Mr. Fizeau and I adopted the idea that the æther is dragged by ponderable matter; and for checking it we have established an experiment producing interference fringes with two beams passing separately through two juxtaposed channels and crossed in opposite directions by two air currents. The experiment undertaken by us in my laboratory has not produced satisfactory results. [...]"*

A note by Fizeau dated October 21, 1850, preserved in the Archives of the Academy of Sciences[113], shows that he had already addressed the problem during the winter of 1848-1849; but he was not using interferometry. Nevertheless, he decided in 1851, visibly inspired by the letter of Fresnel to Arago that we commented on at length above, to undertake an experiment to see what change the velocity of light undergoes in a moving medium. He wrote in the introduction section of the article that describes his experiment[114]:

"Several theories have been proposed to account for the phenomenon of aberration in the system of undulations. Fresnel first, and most recently MM. Doppler, Stokes, Challis, and many others, have published important works on this subject. But it does appear that none of their theories has received the full approval of physicists."

He adds:

*"We owe to M. Arago a method of observation, based on interferences, which is adapted to detect the smallest variations in the refractive indices of a medium. MM. Arago and Fresnel showed the extraordinary sensitivity of this method by several very delicate observations, such as that of the difference in refraction between the dry air and moist air.
An observation mode based on this principle, seemed the only one permitting to measure the change in the velocity [of light] due to the movements."*

Fizeau never made any allusion to any preliminary work that he would have done with Foucault. There is no trace in his notes either, to the point that one may wonder if Foucault, who had just quarreled

---

[112] There is here an allusion to the fact that Arago has been unable to show a variation of the velocity of light linked to the orbital motion of the Earth, and to the fact that the demonstration of Fresnel in 1818 involves aberration (see Box 6.2).
[113] Académie des sciences-Institut de France, Fonds 64.1 Hippolyte Fizeau, dossier 9.26.
[114] *Fizeau H. (1851) *CRAS* 33, p. 349-55. This is the long summary of a Memoir dated 29 September 1851, which was only published in 1859: *Fizeau H. (1859) *Annales de chimie et de physique* 57, p. 385-404.

with him a month before filing his sealed envelope, had not invented the whole story to later claim partial priority: but he never did claim it. However, the principle of the measurement is described there. Fizeau wrote in a note dated the 9th of April 1851, preserved in the Archives of the Academy of Sciences, that he made the first tests of a new device with two copper pipes of 1.50 m in length and 1 cm in diameter (Fig. 6.6). No displacement of the fringes was observed when air was blown with high speed into the tubes.

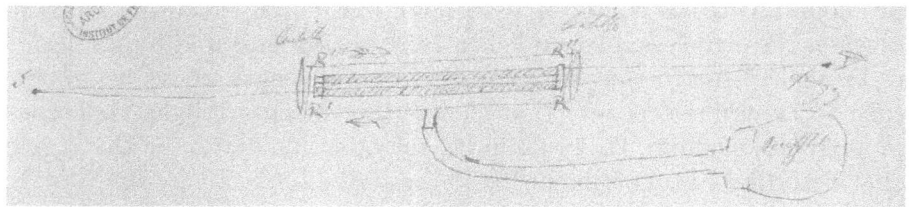

FIGURE 6.6 – Fizeau's autograph diagram of the apparatus for measuring the change of velocity of light in moving air. Driven by a bellow, the air flows from R to R' in the lower tube, and then in the opposite direction R''' to R'' in the upper tube. S is the light source, and interferences between the rays having traversed the tubes are observed at the right. Académie des sciences-Institut de France, Fonds 64.1 Hippolyte Fizeau, dossier 9.02.

Fizeau now mounted a new experiment adapted to the measurement of the variation of the velocity of light in a stream of water. Figure 6.7 is an autograph diagram dated 18th of July 1851 and represents this apparatus.

By circulating water in the tubes with a speed of 7 meters per second, Fizeau measured a displacement of the fringes by 0.23 ± 0.04 fringe. The application of the Fresnel formula for the partial drag of the æther, and therefore of light, by a moving body, predicted a displacement by 0.20 fringe, instead of 0.40 fringe if there was a total drag, that is to say if the velocity of water was simply adding to that of light. Fizeau concluded that Fresnel's formula was correct. For the experiment in air, the calculation of Fresnel predicted a non-measurable displacement by 0.00023 fringe, against 0.41 fringe if the drift was total. Again, the Fresnel formula agreed with the experiment. The velocity v of light in water moving with velocity u was:

$$v = c/n \pm u(1 - 1/n^2),$$

the sign of the second part depending on the direction of the stream with respect to that of the light. c is the speed of light in vacuum and n is the index of refraction of water.

FIGURE 6.7 – Fizeau's autograph diagram of his apparatus for measuring the variation of the velocity of light in a stream of water. The light from a lamp at the lower right was focused on a slit by a lens. A coudé telescope with a 45° semi-reflecting plate then formed a parallel beam. This beam was expanded by two inclined thick plates and was limited by two slits centered on the entrance face of each of the two tubes. The light that has traveled along one of the two tubes was returned to the other by the telescope above, which had a flat mirror at its focus (the same principle as used by Fizeau in Montmartre for his measurement of the velocity of light in 1849, see Fig. 4.3). The interference fringes between the two beams, each of which had travelled the two tubes successively, but in opposite directions, are observed with the eyepiece at the bottom through the 45° semi-transparent plate. The displacement of the central fringe (white light was used), when the water was set in motion according to the arrows, was thus doubled. Académie des sciences-Institut de France, Fonds 64.1 Hippolyte Fizeau, dossier 9.02.

However, Fizeau was not convinced by the explanation of Fresnel. He wrote as a conclusion of his article:

"The success of this experiment seems to me to lead to the adoption of the hypothesis of Fresnel, or at least of the law he found to describe the change of the velocity of light by the effect of the movement of matter; but although this law is verified, giving very strong evidence for the hypothesis of which it is only a consequence, perhaps the conception of Fresnel would seem so extraordinary, and in some ways so difficult to admit, that we still require further evidence, and careful consideration on the part of theoreticians before adopting it as an expression of the reality of things."

How right he was to be skeptical, as will be discussed in the fourth act!

Of course, Fizeau's experiment did not go unnoticed. Despite his poor relationship with Fizeau, Foucault published an honest though succinct report in the *Journal des débats*[115] of the 10th of October 1851, which begins with: "Mr. Fizeau communicates to the Academy the results of work we had undertaken jointly and that he took upon himself to finish alone." But the problems raised by the interpretation of the results were destined to persist for more than half a century.

In addition, although Fizeau has taken all possible precautions to ensure his result, we must admit that the effect was small and that it was not completely immune to systematic errors. That is why, in 1886, Michelson, assisted by his faithful collaborator Edward W. Morley (1838-1923), found it necessary to repeat the experiment. He wrote[116]:

"Notwithstanding the ingenuity displayed in this remarkable contrivance, which is apparently so admirably adopted for eliminating accidental displacement of the fringes by extraneous causes, there seems to be a general doubt concerning the results obtained, or at any rate the interpretation of these results given by Fizeau.

This, together with the fundamental importance of the work, must be our excuse for its repetition."

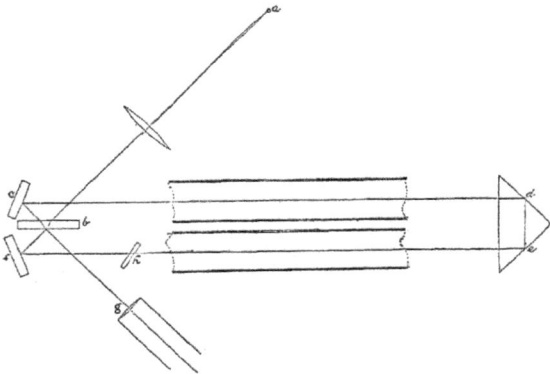

FIGURE 6.8 – Principle of the *refractometer* of Michelson and Morley (1886). The light source was in a, and the parallel beam formed by a lens was separated into two beams by the semi-transparent plate b. One half followed the path $b\ c\ d\ e\ f\ g\ b$ and the other the path $b\ f\ e\ d\ c\ b\ g$. The fringes were observed with the telescope g. Water flowed through the two tubes in opposite directions, as for Fizeau, but now the light passed through each of the tubes in both directions, which made the interferences insensitive to possible fluctuations in refractive index. Wikisource.

---

[115] Accessible via gallica.bnf.fr.
[116] Michelson, A.A. & Morley, E.W. (1886) Influence of motion of the medium on the velocity of light, *American journal of science* 31, p. 377-86, accessible via http://en.wikisource.org/wiki/Influence_of_Motion_of_the_Medium_on_the_Velocity_of_Light

So they built a *refractometer*, whose principle is explained in Figure 6.8. Their experiment fully confirmed the result of Fizeau, this time with a smaller probable error estimated to be only 5%. They also observed that the effect is imperceptible with a stream of air. They conclude that "The result of this work is therefore that the result announced by Fizeau is essentially correct; and that *the luminiferous æther is entirely unaffected by the motion of the matter which it permeates.*" This is a somewhat rough conclusion, since there is still an effect, but less than if the æther was fully swept along.

Pieter Zeeman (1865-1943) took over in 1914, greatly increasing the accuracy of Fizeau's experiment, that he considered as "one of the most ingenious experiments of the whole domain of physics". He achieved this with monochromatic lights of various colors[117]. Chandrashekhara Ventaka Raman (1888-1970) and his collaborator Nihal Karan Sethi (1893-1969) also took this on in 1922, experimenting with air in motion and marginally detecting the very small expected effect[118].

The remarkable experiment of Fizeau is not the only one he has performed to study the drag of the æther in a moving body. After air and water, he considered a solid body:

"To complement and extend the results of the research I have just recalled, it was important to study for the same purpose a solid body such as glass, to see whether the light propagates in it with different velocities when it is at rest or in motion. [...] As for the observation mode, the one that had been previously used for air and water [...] does not work for solids. It was therefore necessary to resort to other principles and use a different method."

This new method was based on the fact that the polarization plane of polarized light passing through an inclined transparent plate, or better a stack of inclined parallel plates, rotates by an angle which depends on the refractive index of the medium of the plates. Fizeau considered that if the plates were moving, the velocity of light within the glass would be slightly modified, which would affect the rotation of the polarization plane. The movement used for this measurement was

---

[117] Zeeman, P. (1914, 1915) Fresnel's coefficient for light of different colours, *Proceedings of the Royal Nœtherlands academy of arts and sciences*, 17, p. 445-51 & 18, p. 398-408, accessible via www.dwc.knaw.nl/DL/publications/PU00012688.pdf and www.dwc.knaw.nl/DL/publications/PU00012515.pdf

[118] Raman, C.V. & Sethi, N.K. (1922) On the convection of light (Fizeau effect) in moving gases, *Philosophical Magazine* 43 p. 447-55, accessible via http://archive.org/details/londonedinburg6431922lond

none other than the orbital motion of the Earth. Fizeau thus illuminated the plates by sunlight reflected by a heliostat and attempted to see if the rotation of the polarization plane varied with the direction of motion of the Earth, thus depending on the time and season. It was not without difficulty, because the effect, if any, was expected to be very small. On the 6th of June 1859, he wrote, full of optimism[119]:

"I believe that God is wanting to reward me for the perseverance that I put in the research of this important phenomenon. The results today do not allow to leave a serious doubt on the reality of the phenomenon I am looking for; the maximum effect is largest around noon and lowest around 4 a.m. as it should be near the solstice for the particular position of the apparatus."

Fizeau was a strong believer[120]. On July 9th, he wrote: "Thanks to God, the demonstration is complete." However, on September 2nd, he was disillusioned: it was not so clear. The notes end inconclusively on October 17th. The article he published on this topic[121] ends with a cautious but optimistic note, while indicating that the experiments will be continued "by the means of a device that will soon be finished." We found no trace of these new measurements, if they have ever been made.

The experiment, had it really given a positive result, would have detected the movement of the Earth with respect to the supposed immobile æther. We now know that this is impossible, and we will see that Fizeau himself helped to show this. But already doubts were emerging. After having reperformed the various experiments that attempted to detect the movement of the Earth with great care, the physicist Éleuthère Mascart (1837-1908) concluded in 1874[122]:

"The general conclusion of this Memoir would be (if we ignore the experiment of Mr. Fizeau on the rotation of the polarization plane by a stack of plates) that the translational movement of the Earth has no appreciable influence on optical phenomena produced with a terrestrial source or with sunlight, and that these phenomena do

---

[119] Académie des sciences-Institut de France, Fonds 64.1 Hippolyte Fizeau, dossier 1: "expériences 1859".

[120] Fizeau succeeded in convincing Le Verrier to receive the sacraments of the Church in June 1877, when he was very ill.

[121] *Fizeau, H. (1859) *CRAS* 49, p. 717-23. The complete article is *Fizeau, H. (1860) *Annales de chimie et de physique* 58, p. 129-63.

[122] Mascart, E. (1874) *Annales scientifiques de l'É. N. S.* 2e série, 3, p. 363-420, see p. 420. Accessible via http://archive.numdam.org

not give us the means to detect the absolute motion of a body: the relative movements are the only ones we can reach."

Did Fizeau's experiment differ from others? Mascart surely had doubts, but Fizeau was alive and respected, and he had to remain cautious. Anyway, Paul Langevin (1872-1946) wrote about the text of Mascart[123]: "This was, enunciated for the first time in final form for optical phenomena, what is now called the principle of relativity, whose subsequent experiments in all domains established the complete exactness [...]".

6.3 Act 3: Fizeau and Michelson again

Fizeau and Mascart were not the only ones trying to demonstrate the motion of the Earth relative to the æther. This is what motivated the famous 1887 experiment of Michelson and Morley. It is less known that Fizeau attempted an experiment for this purpose, in 1852, which he presented in a sealed envelope dated June 9th of that year, with the title: "On the possibility of finding the existence of a relative movement of the æther due to the motion of the Earth by experiment."[124] Here is the beginning:

"The translational movement of the Earth in space must be accompanied in all probability by a current of the luminous Æther that exists in the interior of bodies; this current, when considered in the atmosphere, must occur with a velocity substantially equal to that of the Earth, but in a direction opposite to that of the latter movement. A light source being placed in the atmosphere, the light waves which escape from it propagate in a moving medium. The speed of this medium is 1/10,000 of the velocity of propagation of light.

The waves will be carried by the æthereal medium so that, in the direction of the current of æther, the relative velocity of propagation will be greater than in the opposite direction. It seems certain that it must result from this that the intensity of the light received at equal distances from the light source should not be the same in different directions."

---

[123] Langevin, P. (s.d.) *Annales des Mines*, see http://www.annales.org/archives/x/mascart.html: an excellent scientific biography of Mascart.

[124] The text contained in the sealed envelope and its fac-simile are reproduced by *Acloque (1984), with important commentaries. The paper in *Cosmos* of 1852 is also reproduced by Paul Acloque.

A letter of Fizeau to the editor of the *Cosmos* magazine, Abbé Moigno, specified:

"When the light coming from an incandescent body is received on a screen placed at a distance: the intensity of the received light varying, as we know, as the inverse square of the distance, the intensity increases if the body is getting closer, it decreases if he is leaving.
If the luminous element and the screen are supposed to move while keeping the same distance, the effect of the movement must be the same as if the distance was changing, because the light takes some time to cross this distance, and because the æther in which the light travels does not participate in the motion."

This reasoning is questionable, and the contemporaries and successors of Fizeau offered different formulas for this change in intensity, but this had relatively little importance because the only aim was detection of the phenomenon. Fizeau therefore designed a delicate photometric experiment where he placed two photoelectric detectors back to back at the same distance from identical light sources located on either side, all aligned in the direction of motion of the Earth. He reasoned that the received intensity of the lamp located in the direction opposite to this movement would be larger by $1/2,500$ than the intensity received from the other lamp. There should also have been a very small change in wavelength, that Fizeau that did not mention, but it was irrelevant because the detector used had a very large bandwidth. A galvanometer was to measure the difference in intensity, then the whole set-up had to be turned by 180° to obtain the inverse effect. Fizeau was working on the experiment in June-July 1852 with two different set-ups, each of which had its advantages and disadvantages: one with two light sources and two photocells (Fig. 6.9), the other with a single source and two detectors on each side (Fig. 6.10).

The photoelectric receivers, that Fizeau called *piles*, were thermocouples. These were the only detectors of radiation that were available at that time, apart from the human eye, photography and the thermometer. One of them, which was used by Fizeau, was the *thermomultiplier* of Leopoldo Nobili (1784-1835) perfected by Macedonio Melloni (Fig. 6.11). There are two galvanometers in the Museum of Suresnes that belonged to Fizeau and probably served for his experiments; the best preserved is shown in Figure 6.12.

The drag of æther 91

FIGURE 6.9 – A diagram of Fizeau's attempt to measure the movement of the æther relative to the Earth in 1852. Two identical lamps, probably Carcel lamps with parabolic reflectors, were arranged on either side of a vacuum chamber where two photoelectric receivers were placed at the ends of the central cylinder. They were mounted in opposition and connected to a galvanometer (not shown). The assembly was placed in the direction of the orbital velocity of the Earth and could rotate by 180 degrees around a vertical axis. A primitive version of this set-up is drawn in the sealed envelope deposited by Fizeau at the Academy of Sciences on the 14th of June 1852. Académie des sciences-Institut de France, Fonds 64.1 Hippolyte Fizeau, dossier 9.06.

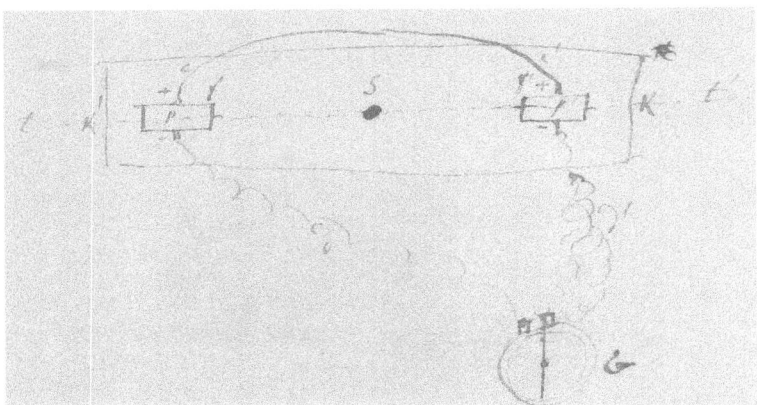

FIGURE 6.10 – Another diagram of Fizeau's from 1852, this time with a single source of light S and two detectors placed symmetrically. These detectors p and p' were mounted in electrical opposition and the resulting current was measured by the galvanometer G. The source and the detectors were aligned in the direction of the orbital velocity of the Earth, and the tray KK' that carried them could rotate by 180° about a vertical axis. This drawing is the draft of the engraving that illustrates the article in *Cosmos*. Académie des sciences-Institut de France, Fonds 64.1 Hippolyte Fizeau, dossier 9.06.

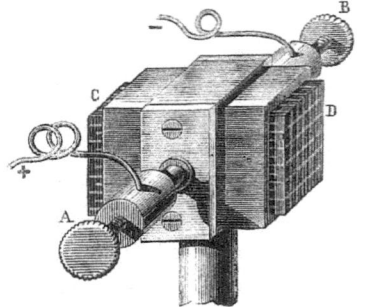

Figure 6.11 – The *thermomultiplier* of Nobili and Melloni, of about 1835. It consisted of 36 antimony-bismuth thermocouples connected in series and separated by thin insulating sheets. The two poles were connected to a galvanometer. The front face was blackened, so that this detector was a bolometer, sensitive to all wavelengths. The name "multiplier" comes from the fact that the detector used multiple thermocouples instead of a single one, increasing its sensitivity. Bibliothèque de l'Observatoire de Paris.

Figure 6.12 – A "multiplying" galvanometer of Nobili, 1835 version, which belonged to Fizeau. A double needle is suspended by a torsion wire. The lower needle, not visible, is located within a flat coil fed by the current to be measured. The upper needle compensates for the terrestrial magnetic field and is used to read the deviation on the circular dial. The whole is protected by a bell jar. The "multiplying" term comes from the fact that the coil was formed of many turns of wire, instead of a simple loop in the first devices. Musée de Suresnes, inv. 997.00.1567.

Of course, Fizeau saw nothing, but the experiment was very difficult given the weakness of the effect expected. He was going to give up, then resumed at the Venteuil Castle where he lived at that time, between 1881 and 1884. The notes kept in the archives of the Academy of Sciences show a stubborn determination to achieve a result. In one of them, dated the 30th of May 1883, he wrote:

"It follows from very numerous and very reliable experiences that I have made on this subject in recent times that the intensity [of light], which in the most favorable direction should vary by 1/5,000, certainly does not change by more than 1/20,000. We must conclude from this that there is necessarily a cause of compensation, which cancels the expected effect. This assumption that seems to me the most plausible is that of an increased intensity in the emission in the direction where the flame shocks the æther and a decrease in the opposite direction [...], compensating exactly for the change of intensity that should take place."

So, Fizeau had not detected anything despite his extreme care (we can imagine the difficulty of measuring differences in light intensity to 1/20,000), and as he still believed in the theory, he thought that another effect was compensating for the expected difference in intensity. He modified his set-up to test this idea: he turned the lamp (or lamps) by 90° and directed their light onto the detector using a 45° mirror, so that the alleged effect of the æther wind on their flame should have disappeared. He thought he saw something and prepared a publication on the 19th of February 1884, but this article never came out: in fact, he continued to make further tests with various modifications, with mixed results, so he was less and less convinced by the results: "the effects of accidental causes are much too large ...". Desperate, he abandoned the problem in June 1884, but he remained obsessed by it until his death.

There was no further mention of the subject, except by a German physicist named Paul Nordmeyer, who, in his thesis of 1903 (see *Acloque 1984), discussed the first experiments of Fizeau (he ignored those of 1881-1884 because Fizeau had not published anything on them). He attempted to redo them, but with little success because the challenges were great: certainly the result was negative, but the dispersion of his measurements was very large.

The famous Michelson and Morley's experiment[125], performed soon after the second series of Fizeau's experiments, had the same purpose: an attempt to detect the relative motion of the Earth and æther. Their interferometric method was very different from that of Fizeau, and unlike the latter, it was of second order: it was sensitive to the square of the ratio of the speed of the Earth to that of light, while the method of Fizeau was directly sensitive to this ratio, hence of first order. The advantage over Fizeau's experiment was clearly apparent, as his measurements were very difficult, while interferometry is so sensitive that the result of Michelson and Morley was certain. We remember only of the latter, but we have to admire the imagination and perseverance of Fizeau. Anyway, both experiments gave a negative result, which was to generate much debate among physicists.

6.4 Act 4: Lorentz, Einstein and von Laue

The impossibility to demonstrate the motion of the Earth relative to the æther was one of the bases of the theory of relativity. The history of the genesis of this theory is too complex to be developed here[126]. One of its steps is the *theory of æther* of Henrik Antoon Lorentz (1853-1928), which is also known under the name *Lorentz's electrodynamics* or *electron theory*[127]. It dates from 1892-5. Lorentz (Fig. 6.13) considered a stationary æther, which represented the absolute reference of Newton and was the carrier of the electromagnetic fields created by the motion of electrons; these fields could only propagate at a velocity less than or equal to the velocity of light, which was assumed to be invariable. Lorentz then explained Michelson and Morley's 1887 experiment by a contraction of the arms of the interferometer which was moving with velocity u with respect to the æther, by a factor $(1-u^2/c^2)^{1/2}$, c being the speed of light; as all standards would be affected by this effect, this contraction was impossible to detect. With his formalism, Lorentz

---

[125] Michelson, A.A. & Morley, E.W. (1887) On the relative motion of the Earth and the luminiferous æther, *American journal of science* 34, p. 333-45, accessible via http://www.aip.org/history/gap/PDF/michelson.pdf
[126] See in particular Darrigol, O. (2004), http://www.academie-sciences.fr/activite/archive/dossiers/Einstein/Einstein_pdf/Darrigol%20_amp.pdf
[127] Lorentz, H.A. (1895) *Electrischen und optischen Erscheinungen in bewegten Körpen*, Leiden, E.J. Brill, accessible via http://de.wikisource.org; Lorentz, H.A. (1916) *The theory of electrons and its applications to the phenomena of light and radiant heat*, Leipzig, Teubner, see p. 190-1, accessible via http://archive.org/details/electronstheory00lorerich

could also explain the aberration of starlight, the Doppler-Fizeau effect and finally the change in the velocity of light in moving fluids, without resorting to the partial drag of the æther invoked by Fresnel.

FIGURE 6.13 – H.A. Lorentz in 1902. Official photograph of the Nobel prize, Wikimedia Commons.

However, the final explanation was only possible after Albert Einstein had published his famous article *On the Electrodynamics of Moving Bodies* in 1905. He provided the relativistic formula for the composition of velocities in this article:

$$v = (u + w)(1 + uw/c^2)^{-1},$$

where u and w are the two velocities that compose themselves, v the resulting velocity and c the velocity of light

If any of these velocities, here w, is that of light, this equation becomes:

$$v = (u + c)(1 + u/c)^{-1} = c,$$

which shows that the velocity of light cannot be exceeded, and that it is not possible to measure the velocity of the Earth by observing the light from an external source, as Arago, Fizeau, Michelson and many others had hoped.

Two years after Einstein's publication, Max von Laue (1879-1960) published a paper[128] in which he used the relativistic composition of velocities to find Fresnel's formula. Box 6.3 reproduces his demonstration.

---

[128] von Laue, M. (1907) *Annalen der Physik* 23, p. 989-90.

Box 6.3. Relativistic demonstration of Fresnel's formula

Let us consider Fizeau's experiment on the velocity of light in a stream of water. The velocity of light in the moving water, relative to this water, is c/n, n being the refraction index of water. Relative to a fixed observer, the velocity v of light in water that moves with a speed u is given by Einstein's equation for the composition of velocities:

$$v = (c/n + u)(1 + u/nc)^{-1}$$

As u/nc is much smaller than 1, we can make a limited development of the numerator, and the equation becomes, limiting ourselves to the terms of the first order with respect to u:

$$v = (c/n + u)(1 - u/nc + ...) = c/n + u(1 - 1/n^2) + ...$$

which is Fresnel's equation.

Von Laue noted that the problem was solved much more simply using the theory of relativity than using Lorentz' theory. This ended nearly a century of questions and uncertainties. And he concluded: "In this way we do not need to introduce an "æther" that penetrates the body without sharing its movement." This was the beginning of the death of the æther, which would however die hard, since Einstein has not yet abandoned the æther when he produced the theory of general relativity in 1915; it still remained in the minds of many physicists for some years.

Chapter 7
The diameter of stars

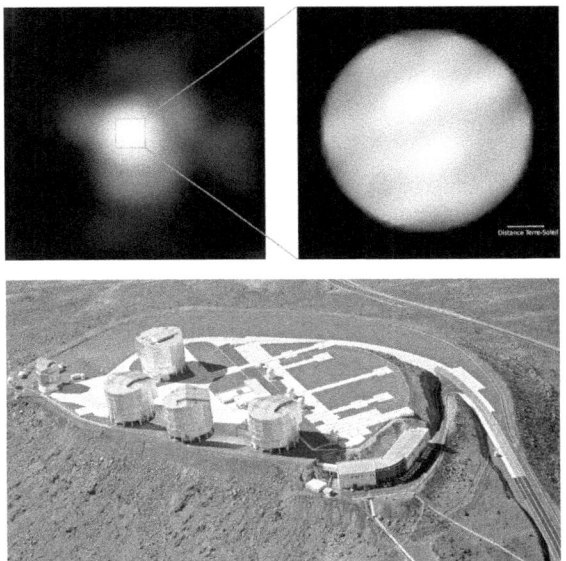

Top, two images of the supergiant star Betelgeuse (α Orionis). On the left, in the infrared from one of the 8-m diameter telescopes of the Very Large Telescope (VLT) of the European Southern Observatory, obtained with adaptive optics (real-time correction of the effect of atmospheric turbulence by a deformable mirror). The star is not resolved, but we observe the gas it ejects. Right, Betelgeuse is resolved in this infrared image obtained by interferometry with the Infrared Optical Telescope Array (IOTA) in Arizona, with an angular resolution of 0.009 seconds of degree. This is one of the first existing images of a star other than the Sun. (*From Haubois, X. et al. (2009) Astronomy & Astrophysics, 508, 923-932, with the permission of ESO*).

Bottom, the interferometer of the Very Large Telescope of ESO (VLTI) with its 4 small telescopes of 1.8 m in diameter, movable on the white areas, which can be combined with the 4 large fixed 8-m telescopes. The assembly can operate as a multi-element infrared interferometer with bases up to 200 m, in order to produce images of celestial objects with an angular resolution of about a millisecond of a degree. This is the ultimate development of the idea of Fizeau. ESO.

7.1 A brilliant idea

Amongst Fizeau's manuscripts preserved in the Archives of the Academy of Sciences, there is a text from 1851 entitled "On a way to derive the diameters of the stars from some interference phenomena."[129] It was an entirely new idea that, long after, would lead to an enormous development. Yet Fizeau never published it, except in 1868 as a simple note in a report to the Academy of Sciences[130]:

> *"There exists [...], for most of phenomena of interference, such as Young's fringes, Fresnel's mirrors and those that give rise to the scintillation of stars according to Arago, a remarkable and necessary relationship between the size of the fringes and that of the light source, so that fringes of extreme thinness can occur only when the light source has almost imperceptible angular dimensions; hence, to say it in passing, there may be some hope that based on this principle and forming for example, using two very wide-spaced slits, interference fringes at the focus of the large instruments used to observe the stars, it will be possible to get some new information on the angular diameters of these stars."*

Indeed, nothing was known at that time about the dimensions of the stars and only a few stellar distances were known; for lack of anything better, it was often assumed that all stars were similar to the Sun when one wanted to estimate, for example, the size of the Milky Way, as William Herschel did in 1785. It would have been important to measure at least their angular diameter, and this is what Fizeau proposed in 1851. We translate his manuscript in full below, because of its importance:

> *"June 22, 1851 (important)*
> *Application of interference phenomena to the measurement of very small angles such as the angles subtended by the rays emanating from the two opposite edges of the star – (diameter of the stars) –*
> *In telescopes, the images of stars always have a sensible diameter that is quite accidental and is due to the imperfection of mirrors and lenses that do not focus exactly to the same point the incident*

---

[129] Académie des sciences-Institut de France, Fonds 64 J Hippolyte Fizeau, dossier 9.01.
[130] *Fizeau, H. (1868) *CRAS* 66, p. 932-4.

parallel rays[131]; this circumstance precludes that we can assess and measure the very small diameters of the images of stars. For, if the actual diameter is significantly smaller than the <u>accidental</u> [emphasized by Fizeau] apparent diameter due to the imperfection of the instrument, this real diameter could decrease indefinitely without any change in the apparent diameter. The diameters of stars appear to be in this case, their actual size not being measurable with the best instruments.

It seems to me that the interference phenomena can lead to something new on this subject, by allowing to estimate much smaller angles than those observed at the focus of telescopes.

In most circumstances where one observes the mutual influence of light rays, such as in Fresnel's mirrors or Young's slits, either alone or associated to a converging lens as I mentioned several years ago, we see that the light source must be very small for fringes to occur. The size of the light source is related to the width of the fringes that are produced, and for each particular width there is a size limit for the source that cannot be exceeded without disturbing and destroying the fringes. This phenomenon is easily explained by the superimposition of the different fringe systems produced by each point of the light source. I am just recalling the existence of this explanation without trying to give it completely. It always seemed very striking to me to recognize, through the inspection of the fringes, if the light source has a diameter of 1/20, 1/50 or 1/100 of a millimeter, and to estimate in this way very small angles subtended by the rays sent by the two opposite edges of the source. These angles are usually 20", 10" and even 1"."

Fizeau performed the very simple calculation of the angular distance between the fringes in a Young's experiment where the slits are separated by d: it is simply λ/d, λ being the wavelength. If the light source, a star for example, has an apparent diameter of this order, the fringes produced by the various points of this source are shifted and their superposition is such that the contrast of the observed fringes becomes very small: we cannot see them. Figure 7.1 reproduces a handwritten page of Fizeau's, undated, where he made these calculations.

---

[131] In fact, the spread of stellar images is mainly due to the effect of atmospheric turbulence, as it also exists with high-quality telescopes. Fizeau, who was not an astronomer, seemed to ignore it.

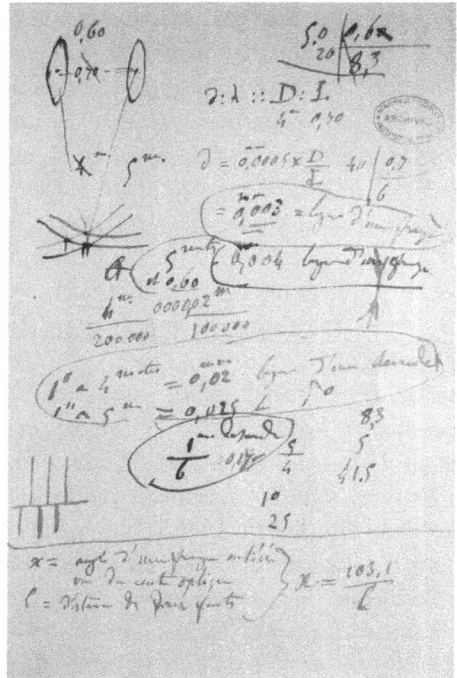

FIGURE 7.1 – Fizeau's manuscript, undated but certainly later than June 22nd, 1851, containing the calculation of the dimensions of the interference fringes produced by the two crescents represented at the upper left. Académie des sciences-Institut de France, Fonds 64 J, Hippolyte Fizeau, dossier 9.01.

Here is the remaining part of the manuscript:

"By placing two slits in front the lens of a telescope with 80 cm focus and looking at the image of a light line formed at the focus of a cylindrical lens, I could distinguish from the inspection of the fringes if the light source subtended an angle larger than 2" or even 1", while in observing with the telescope itself the image of the source, angles 10 times larger could not be distinguished from these small angles, the image keeping substantially the same diameter despite the change in the size of the source.

This remark leads to think that we can produce interference fringes with starlight in conditions such that the light source subtends an angle of less than 1" or 1/10" in order to see fringes. We could decide whether a given star actually has a diameter of this order of magnitude while a good telescope does not allow to measure such small diameters, which all result in images of the same size.

I assume that the angular diameter of the source being equal to that of a fringe (the central one) the fringes must be invisible for larger diameters and visible for smaller diameters. [In the margin: it should be the double, from experience].

So I suppose a telescope followed by a device for producing fringes, in conditions such that these fringes should disappear for a source diameter greater than 1". By directing the instrument to different stars we could decide immediately which ones have angular diameters larger than 1", and if such stars exist. For them, indeed, fringes would not occur while conversely, for any other, fringes would be visible whatever their diameter.

The apparatus being arranged to cease to obtain fringes for a diameter of 1/10", we could recognize all the stars whose diameter is greater or less than this new limit.

Thus, by varying the limit, we could appreciate with some accuracy these small angles that it does not seem possible to measure with the known instruments.

For light points placed on the surface of the Earth or at a small distance the success of this method makes no doubt; for stars, it is to be feared that the changes in density, temperature, humidity that scintillation produces in the layers of air, even the neighboring ones, precludes the production of fairly clear and unchanging fringes in this kind of observations; however we have to test that, the thing is worth it.

1. Do an experiment with two circular openings placed before the lens of a telescope."

Fizeau accompanies this sentence with a little drawing on the margin, which is the same as that at the top left of Figure 7.1 but without the numbers, with the indication "most advantageous form."

"If the fringes are relatively steady in calm nights or if there is little scintillation we can use the following set-up which would give a much greater sensitivity [it is represented Fig. 7.2].

2. 4 mirrors m m' m" m"" parallel two by two, two at the center forming an angle of 90°. One would thus obtain in O fringes whose separation would correspond to the distance of the slits ff'; and they would be as separated as wished.

The <u>limit angle</u> [that is to say the resolution] would depend on the distance m m''', or more accurately mm' + m"m"'.[132] It would be substantially equal to the angle subtended by the central fringe

---

[132] This is wrong: only the distance $m\ m'''$ intervenes in the problem.

formed in O by light coming directly from m and m'''. Thus it would be as small as one would like.

Note: *For [crossed out, illegible], we could use total reflection prisms because we will never have enough light [drawing Fig. 7.3]."*

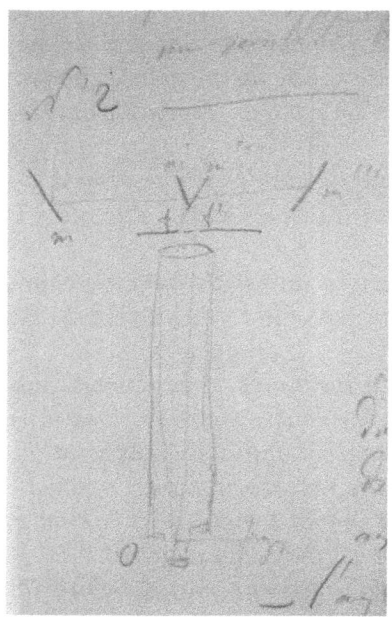

FIGURE 7.2 – A set-up to measure the diameter of stars. See text. Académie des sciences-Institut de France, Fonds 64 J, Hippolyte Fizeau, dossier 9.01.

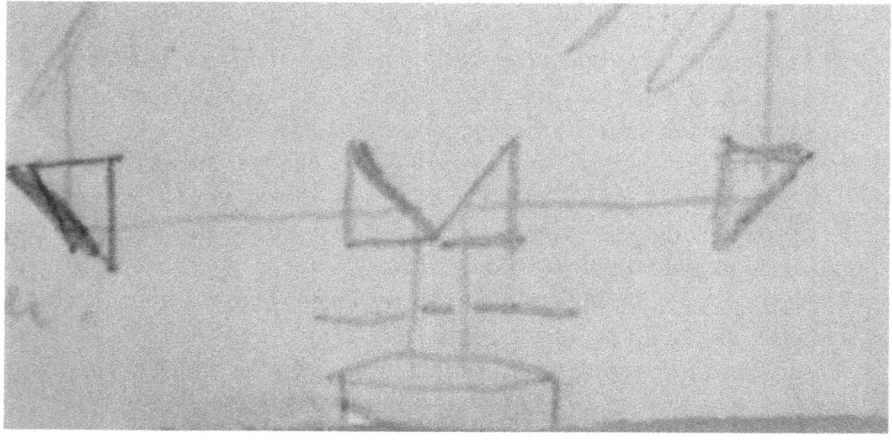

FIGURE 7.3 – Another set-up to measure the diameter of stars. See text. Académie des sciences-Institut de France, Fonds 64 J, Hippolyte Fizeau, dossier 9.01.

On the title page of the file, we find the design reproduced in Figure 7.4 with the following annotation, apparently added later: "The most practical arrangement would be: two achromatic lenses and two prisms".

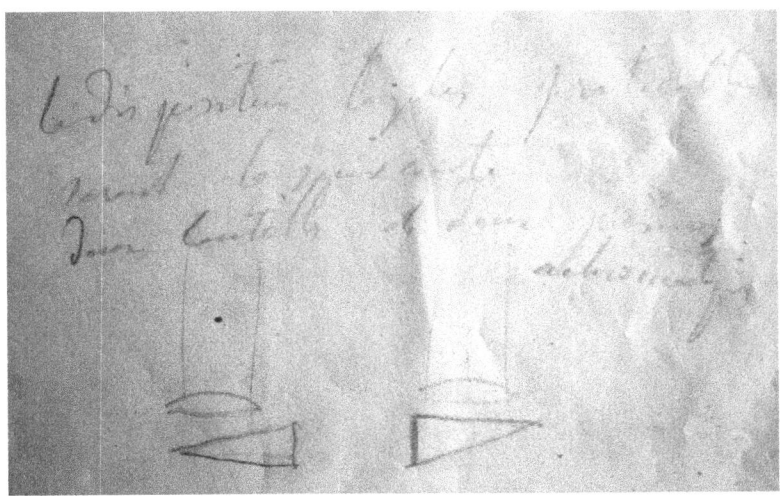

FIGURE 7.4 – Still another set-up to measure the diameter of stars. See the text. Académie des sciences-Institut de France, Fonds 64 J, Hippolyte Fizeau, dossier 9.01.

Although Fizeau's text is somewhat convoluted, it is clear that he understood perfectly well that we can estimate the diameter of stars by interferometry and that the resolving power is better if the distance between the two openings which produce fringes is larger. He does not appear to have put his idea into execution itself, which is a shame, because, even without a small astronomical instrument, he could have used the arrangement of Figure 7.4 placed on a good equatorial mount. The arrangement of Figure 7.1 was used in 1873-4, as we will see, and we recognize the principle of the set-up used by Michelson and Pease in 1921 in Figure 7.2.

7.2 The first tests

The idea of Fizeau had not gone unnoticed, even though he did not publish it before 1868. We do not know how Édouard Stefan (Fig. 7.5, Box 7.1), who had been running the Marseilles observatory since 1866, learnt of the existence of this idea and came into contact with Fizeau. In any case, it is clearly Fizeau who suggested he measure the apparent diameter of stars by interferometry. Stefan had the largest telescope of the time in Marseilles, the Foucault reflecting telescope of 80 cm in diameter (Fig. 7.6),

FIGURE 7.5 – Édouard Stephan. Wikimedia Commons

FIGURE 7.6 – The 80-cm reflecting telescope of the Marseilles observatory, old photograph. Built by Foucault and installed in 1866 in a beautiful dome also designed by Foucault, this telescope remained the largest in the world until 1895. It has been used for a century and is still visible in Marseilles, but the dome has been destroyed. Bibliothèque de l'Observatoire de Paris.

Box 7.1. Édouard Stephan (1837-1923)

A former pupil of École Normale Supérieure, major in mathematics, Stephan was noticed by Le Verrier in 1866, who appointed him as director of the Marseilles observatory that he had created as a "branch of the Paris Observatory." He lead it until 1907, and had a brilliant observer career with the telescope of 80 cm in diameter that Foucault had installed in 1864, which remained, for a long time, the largest in the world. In 1866, Stephan participated in an expedition to Siam to observe a total solar eclipse, where he discovered, with Georges Rayet (1839-1906), a new line in the spectrum of a protuberance. It was later identified as a line of helium, seen in the Sun before being isolated on Earth. In 1873-4, he tried to measure the apparent diameter of many stars with interferometry, to find that it was always less than 0.16 seconds of a degree. He discovered no less than 800 "nebulae" that proved to be almost all galaxies.

In a letter to Fizeau[133], Stephan described his observations. We translate it here in full:

"Marseilles, 1874 February 1
Sir,
The length of the silence I kept towards you, might have seemed excessive; you have not assigned it, I hope, to my negligence; because I would have been really guilty of neglect or indifference in the interesting study of the diameter of the fixed stars. On the contrary, despite important multiple occupations, and despite the considerable time taken up by my observations of nebulae, the first hundred of which I have just completed, I have not stopped a single day, since the time when I had the honor to see you for the last time, to worry about the important and new issue that I started under your auspices, taking profit as much as I could of the advice you have given to me with such perfect kindness.
I was refraining from writing to you until I would finally encounter a star that would not produce fringes, and patiently pursuing my

---

[133] Académie des sciences-Institut de France, Fonds 64 J Hippolyte Fizeau, dossier 11.40.

explorations, I thus delayed my letter from week to week, without finally getting the result we were both hoping for.

After my return from Paris, I have sought to standardize and improve the experimental conditions. After various tests, I decided for a screen with two crescents placed directly on the mirror. It is with this disposition that the flexures of the telescope have the least influence. Now, this is capital; because, for a fringe, it is necessary that the two beams received in the microscope eyepiece keep nearly the same intensity and it is quite difficult to adjust the relative positions of the mirror, of the screen and the total reflection prism [which sends the beam to the side in the Newton mounting of the telescope] so that one of the beams does not acquire a more or less great preponderance when the instrument is tilted.

The screen that I use today is pierced by two crescent-shape openings limited by equal circles, 80 centimeters in diameter; the major axes of these crescents are parallel and their distance is 0.65 m [drawing reproduced Fig. 7.7].

One can hardly exceed this spacing: beyond, the images will weaken in an exaggerated manner and lose too much of their sharpness. This drawback arises from the fact that, in all the telescope mirrors, regardless of the quality of the work, the periphery is somewhat less perfect than the rest of the surface. The eyepieces are nine in number, the respective magnifications being 150, 200, 300, 400, 500, 600, 900, 1200, 1500 times.

For nine months I have observed most of the visible stars, including those of the 3rd magnitude and some of the 4th. <u>All gave me fringes</u>. You can imagine how eagerly I awaited the time when Sirius could be seen at the meridian in favorable circumstances, for an initial examination performed under quite defective conditions had led me, last year, to suspect an exception for this star. For a month and a half I could look at it at my ease; <u>Sirius, like the other stars, gave very sharp fringes</u>.

In the first moment, I confess, I had a great disappointment, a feeling against which I hastened to respond, because we should not have a bad mood against the facts if we wish to study them well.

Moreover, is the result less interesting than the opposite one? I do not believe that. Conversely, it emerges from these observations an important consequence.

Indeed, it is quite remarkable that the appearance of fringes begins to manifest itself in much the same way for all the stars, at least with the same magnification. The difference can be large for different observers, but as for me, I have always begun to discern the fringes with a magnification of 600 times, when the undulations of the images were not too strong. It must be concluded that the superimposition of the fringe systems produced by extreme waves [from the edges of the star] is very small compared to the spacing of the fringes of each system; in other words, the apparent diameter of the star seems insignificant compared to the apparent diameter of the fringes seen from the center of the objective.

Thus, the apparent diameter of all observed stars is considerably less than 1/6 of an arc second.

If I am not mistaken, this is a well established concept, the first that has been obtained on the matter. Such a result is not without importance. Moreover, it undermines in no way the hope that we had to determine the diameter of some stars. The principle of the method remains, the instrument is too small, that's all. It will be very interesting to see what will give the 1.20-m telescope currently under construction in Paris.

I confine myself today, Sir, to some thoughts; if you think there is some utility to this, I will send you shortly a note more technical that the present one, with an analytical development of the problem by the method of Mr. Kochenhauer.

Please accept, Sir, the assurance of my most respectful sympathy. E. Stephan"

FIGURE 7.7 – Drawing of the diaphragm placed by Stephan on the mirror of the 80-cm telescope to obtain the fringes. Compare with Figure 7.1. Académie des sciences-Institut de France, Fonds 64 J, Hippolyte Fizeau, dossier 11.40.

Why Sirius? Because, Sirius was the brightest star in the sky, it was believed (wrongly) that it should have the largest apparent diameter. In a first step, Stephan believed to have resolved Sirius, which seemed not to give fringes[134]. But finally, he found that all the stars he observed gave fringes, including Sirius, so their apparent diameter must have been significantly less than 0.16 seconds of a degree[135]. This was a very important result, although no star was resolved. Stephan hoped that the telescope of 120 cm in diameter, then under construction at the Paris Observatory, would resolve some stars by interferometry, but its optical quality was so bad that it could not be used for this purpose. Anyway, it would still have been too small.

7.3 Michelson again!

In 1890 and 1891, Michelson published two articles[136,137] where he explained that we can accurately measure the apparent diameter of the stars by placing two slits in front of the lens of a telescope and examining the interference fringes thus produced. This was the same as Fizeau's idea, which he cited nowhere, however. As Michelson had much admired Fizeau's experiments on the velocity of light in moving water, it is possible that he ignored his remark of 1868 and the results of Stephan, which he did not cite either. Anyway, his articles indicated the principle and contained comprehensive calculations. He introduced, in particular, the function of visibility of the fringes, which gives their intensity as a function of the spacing of the slits for multiple observations of a given source, in this case a circular and uniform source as the disk of a star: it shows a decreasing series of maxima separated by periods in which the fringes disappear completely (Fig 7.8). The first zero occurs when the apparent diameter of the star is 1.22 λ/d radians, λ being the wavelength and d the spacing of the slits.

---

[134] *Stephan (1873) *CRAS* 76, p. 1008-10.
[135] *Stephan (1874) *CRAS* 78, p. 1008-12.
[136] Michelson, A.A. (1890) On the Application of Interference Methods to Astronomical Measurements, *Philosophical Magazine*, 5th series 30, p. 1-20, accessible via http://www.biodiversitylibrary.org/item/122067
[137] Michelson, A.A. (1891) Visibility of Interference-Fringes in the Focus of a Telescope, *Philosophical Magazine*, 5th series, 31, p. 256-9, accessible via http://www.biodiversitylibrary.org/item/122067; see also °*Publications of the Astronomical Society of the Pacific* 4, p. 217-20.

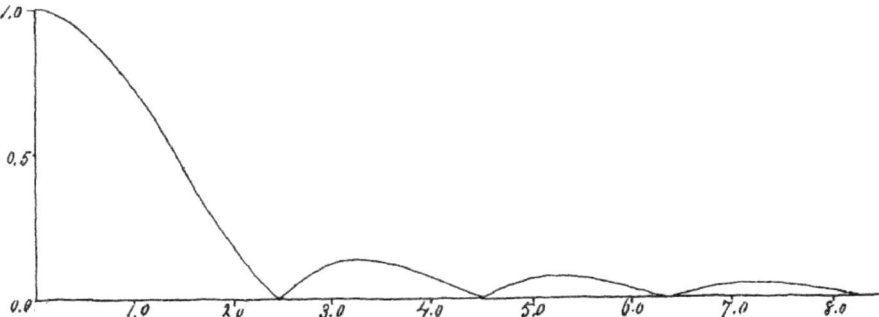

FIGURE 7.8 – The visibility function of the interference fringes produced by two slits. The intensity of the fringes is given in ordinates as a function of the slit spacing, for a source which is a uniform circular disk. In abscissæ, the ratio between the angular radius of the object (in radians) and λ/d, where λ is the wavelength and d the distance between the slits. From Michelson (1891).

As a test, Michelson applied this method to measure the diameter of Jupiter's satellites, which was pretty well known by measurements with eyepiece micrometers[138]. He used the 30-cm diameter telescope of the Lick Observatory in California and obtained the results shown in Table 7.1.

Slightly later, in 1895, Karl Schwarzschild (1873-1916) measured, by a similar method, the spacing between the components of double stars in Munich[139]. In 1898, Maurice Hamy (1861-1936), who extended the calculations of Michelson, made similar measurements on the satellites of Jupiter with the Grand Equatorial Coudé of the Paris Observatory, whose diameter was 60 cm[140]. His results are compared with those of Michelson in Table 7.1. In reality, neither were better than direct measurements with a good ocular micrometer for those objects, whose apparent diameter is of the order of a second of a degree,.

---

[138] °Michelson, A.A. (1891) Measurements of Jupiter's Satellites by Interference, *Publications of the Astronomical Society of the Pacific* 4, p. 274-8.
[139] °Schwarzschild, K. (1895) Ueber Messung von Doppelsternen durch Interferenzen, *Astronomische Nachrichten* 139, p. 353-60.
[140] °Hamy, M. (1899) *Bulletin astronomique* 16, 257-73.

Table 7.1 : Diameters of the satellites of Jupiter and the asteroid Vesta measured by Michelson and by Hamy.

	I: Io	II: Europe	III: Ganymède	IV: Callisto	Vesta
Michelson, 1891	3850 km	3550 km	5170 km	4940 km	-
Hamy, 1899	3550 km	3150 km	4640 km	5100 km	392 km
Modern value	3630 km	3138 km	5268 km	4800 km	576 km (non spherical)

Then nothing happened until, in 1919, Michelson and his partner Francis G. Pease (1881-1938) received funding for measuring the diameter of the star Betelgeuse (α Orionis) by interferometry. The knowledge of stars had improved considerably since the previous century. In 1900, Max Planck (1858-1947) published his theory of blackbody radiation, soon verified by laboratory experiments. If we can determine the surface temperature of a star and if it is assimilated to a black body, this theory allows the calculation of its brightness per unit area in the visible light spectrum, then, knowing this brightness, to estimate its apparent diameter from its flux. However, only a few stellar temperatures were well determined in 1919, but not that of Betelgeuse[141]. In 1914, William W. Coblentz (1873-1962) had measured the infrared radiation of a number of stars, including M-type cool stars (like Betelgeuse), and found that most of their radiation was in the infrared[142]. For lack of measurements in the infrared, it was hard to predict what might be the apparent diameter of Betelgeuse, even if it was obvious that it had to be significantly larger than that of Sirius, a hot star. This was the motivation of the successful interferometric measurement made in 1920 by Michelson and Pease with the 2.50-m diameter telescope of Mount Wilson in California[143] (Fig. 7.9): they found, by observing the first cancellation of the fringes, a diameter of 0.047 seconds of a degree. The distance of Betelgeuse was still poorly determined, but it was already clear that it was a huge star, a supergiant, its diameter being comparable to the orbit of Mars. Pease determined the apparent diameter of Arcturus

---

[141] See °Lequeux, J. (2010) Charles Nordmann and multicolour stellar photometry, *Journal of Astronomical History and Heritage* 13, p. 207-219.
[142] °Coblentz, W.W. (1914) Note on the radiation from stars, *Publications of the Astronomical Society of the Pacific* 26, p. 169-178
[143] °Michelson, A.A. & Pease, F.G. (1921) Measurement of the diameter of α Orionis with the interferometer, *Astrophysical Journal* 53, p. 249-59.

(α Bootis) with the same apparatus, but the length of the base of the interferometer was still insufficient to resolve other stars[144]. The 2.50-m telescope was also used by John A. Anderson (1876-1959) to study double stars with two variable-spacing slits, extending the measurements of Schwarzschild[145].

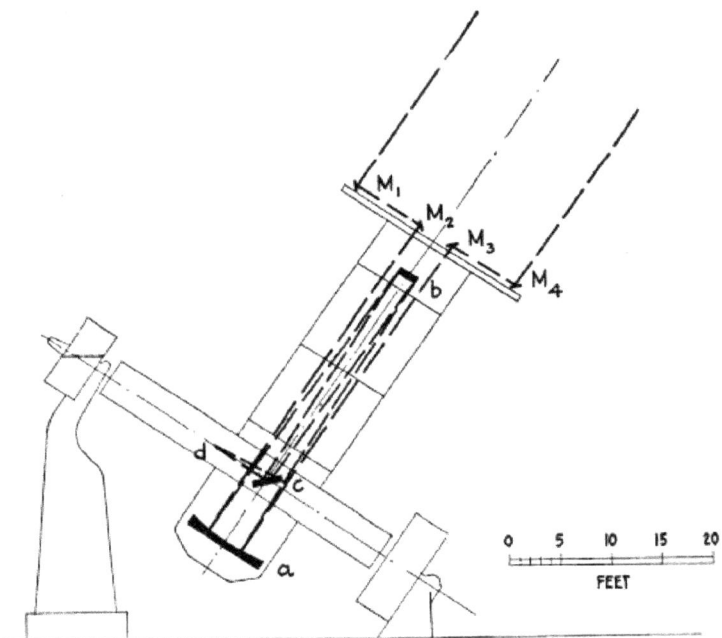

FIGURE 7.9 – Diagram of the apparatus used on the 2.50-m diameter telescope of Mount Wilson to measure the diameter of Betelgeuse. The 45° mirrors M_1 and M_4, whose spacing was adjustable, defined the basis of the interferometer. They reflected the light from the star to the mirrors M_2 and M_3, which sent it to the primary mirror of the telescope. The beams were then returned to the convex secondary mirror b, then to the 45° mirror c that directed the beam along the polar axis of the telescope. The fringes were observed in d. Compare with Figure 7.2. From Michelson and Pease (1921), with permission from the American Astronomical Society.

---

[144] °Pease, F.G. (1921) The angular diameter of α Bootis by the interferometer, *Publications of the Astronomical Society of the Pacific* 33, p. 171-3.
[145] °Anderson, J.A. (1920) The Michelson interferometer method for measuring close double stars, *Publications of the Astronomical Society of the Pacific* 32, p. 58-9.

7.4 Stagnation and renewal of astronomical interferometry

After the success of the measurement of the diameter of Betelgeuse, Pease built an interferometer with a larger base to measure the diameter of smaller stars. Michelson died in 1931, the year of the completion of this interferometer. Pease did not get any good results with this instrument, which he abandoned in 1938. This was going to discourage astronomers who wanted to continue this work, until a new interferometric technique appeared in 1956.

Meanwhile, interferometry had made tremendous progress in the study of the structure of celestial sources of radio waves, that is to say in radio astronomy. Sophisticated interferometric techniques now made it possible to use many antennae and to obtain real images with high angular resolution. It was a British radio astronomer, Robert Hanbury Brown (1916-2002) who, with his compatriot, the great amateur physicist Richard Q. Twiss (1925-2005), took over optical interferometry using a technique developed by radioastronomers: intensity interferometry. This time, it was enough to detect the signal from a star with two telescopes, and to correlate the signals from the two detectors (photomultipliers). Although physicists initially had trouble understanding what was happening (which is explained by the Bose-Einstein correlation of photons, which are bosons[146]), it worked well and our two scientists succeeded in measuring the apparent diameter of Sirius[147]. This encouraged them to build a large intensity interferometer in Australia (Fig. 7.10) with which they measured the apparent diameter of 32 stars with great accuracy[148].

However, this technique lacked sensitivity, and the authors themselves consider that it had no future. We had to wait until 1974, when Antoine Labeyrie in France resumed the interferometric technique of Michelson. This time, he used two small telescopes (Fig. 7.11) between which the interference fringes were produced, the optical paths were equalized by moving the table where the fringes formed[149]. Presently, this equalization is generally done by optical delay lines.

---

[146] See for example http://en.wikipedia.org/wiki/Hanbury_Brown_and_Twiss_effect
[147] °Hanbury Brown, R. & Twiss, R.Q. (1956) A Test of a New Type of Stellar Interferometer on Sirius, *Nature* 178, 1046–8
[148] °Hanbury Brown, R., Davis, J. & Allen, L.R. (1974) The Stellar interferometer at Narrabri Observatory, *Monthly Notices of the Royal Astronomical Society* 137, p. 375-92
[149] °Labeyrie, A. (1975) Interference fringes obtained on Vega with two optical telescopes. *Astrophysical Journal* 196, L71-75.

FIGURE 7.10 – The intensity interferometer of Narrabri, Australia. It consisted of two telescopes with composite parabolic mirrors of 6.5 m in diameter, movable on a circular track, 188 m in diameter. The light from the star was focused on a photocell (photomultiplier) carried by the long pole. Photograph by Prof. John Davis (1932-2010), University of Sydney.

FIGURE 7.11 – The first interferometer of Antoine Labeyrie, at the Nice observatory in 1975. The two small telescopes were spaced by 12 m on a north-south fixed base, and directed the light of the observed star to the central shelter, where interference fringes were formed on the superposed pupils of the two beams. The observation was made during the passage of the star at the meridian. From Labeyrie, 1975, with permission from the American Astronomical Society.

Since then, optical interferometric techniques have developed enormously, and at least ten interferometers with several mobile telescopes have been built[150]. They allow real two-dimensional images to be obtained. They generally work in the near or mid infrared, where technical difficulties are smaller than in visible light because the atmospheric turbulence is less troublesome. One of them is the Very Large Telescope Interferometer (VLTI) of the European Southern Observatory, shown at the beginning of this chapter. The idea of Fizeau, continued by his intellectual successors Michelson and Labeyrie, has produced incomparable fruits!

---

[150] For a fairly comprehensive study until 2006, see Lawson (2006). The history of interferometry in France until 2005 is summarized in Lequeux (2005), Appendix 1.

Chapter 8
A highly esteemed scientist

Photograph and signature of Hippolyte Fizeau. Académie des sciences-Institut de France.

8.1 A shortened family life

On the 1st of September 1853, Fizeau married Thérèse Valentine de Jussieu, under community property. She descended from the famous naturalists Antoine Laurent de Jussieu (1748-1836) and Pierre Bernard de Jussieu (1751-1836) (see the family tree of Fizeau and his wife in Appendix 2). Before her marriage, Thérèse Valentine was staying with her grandmother, Thérèse Adrienne Jussieu, 61 rue Cuvier, at the Museum of Natural History. As for Fizeau, he was living in Paris, 4 rue Garancière. After 1860, Fizeau spent more and more time at the castle of the Jussieu family, the Venteuil castle, a beautiful eighteenth-century building located in Jouarre (Seine et Marne) (Fig. 8.1). It was easy to go there because the railway had connected Paris and La Ferté-sous-Jouarre, not far from the castle, since 1849; in 1862 it was considered as his main residence. Fizeau then performed all of his experiments there; much of the material he used and some of his notes were given to the Urban and Social History Museum of Suresnes by a descendant of the sister of Fizeau's wife, Bernard Ramond Gontaud (1903-1967). Fizeau still retained a home base at 3 rue Vieille-Estrapade, later called rue de l'Estrapade, on the Montagne Sainte-Geneviève in the Latin Quarter. Indeed, he went to Paris every week to attend the meeting of the Academy of Sciences, of which he was one of the most active members, and also, after 1878, the meetings of the Board of longitudes.

In 1881 he bought a house for 100,000 francs, located at 21 Quai de l'Horloge in Paris[151]. On his death in 1896, his fortune was estimated at 1,167,814.90 francs, plus half of the Venteuil castle (estimated to be 521,840 francs in 1934, including the adjoining land).

Fizeau had four children, one son and three daughters (see Appendix 2). The first three children were born in Paris, and the last one in Jouarre. The first daughter, Thérèse Zoé, died at the age of 7 in 1866, causing great sorrow to Hippolyte, who had already lost his wife on the 3rd of September 1863. He then took refuge by immersing himself in science. His family woes affected his character

---

[151] One may estimate the value of the French franc at about 4 dollars during the 19th century.

and his lifestyle so much that his pupil Cornu portrayed him as follows in a scientific leaflet published just after his death[152]:

Fizeau was a reserved character, cautious, a little touchy: he confided rarely in others, but always rightly. He had some hidden enthusiasm, and his word, usually so calm, became animated with a singular vivacity when he expressed his somewhat mystical admiration for great things or for the great geniuses who had accomplished them. [... Since the death of his wife,] he lived in isolation with his three children, leaving his retreat only to attend the meetings of the Academy and the Bureau des Longitudes: it was the only distraction he allowed to himself. He had also the pain of seeing his eldest daughter die: to withstand this new ordeal he began a long and hard research that he could unfortunately not complete.

The "long and hard research" to which Cornu alludes was of course his attempt to detect the movement of the Earth relative to the æther using photometry, to which we devoted a part of Chapter 6. He appears to have abandoned it in 1884, eight years before his death. But he had spent so much time in vain on this research that he probably remained obsessed by the problem, and he might have told Cornu of his intention to resume it some day...

FIGURE 8.1 – The Venteuil castle in Jouarre (Seine et Marne), aerial view from the north. Built around 1760 by the architect Anne Alexander Saget des Louvières, who constructed many buildings in the area, it is surrounded by a park with a terrace that opens to the north, overlooking the Marne valley. Acquired in 1810 by Antoine Laurent de Jussieu, the grandfather of Fizeau's wife, it was the main residence of the physicist from the 1860s until his death in 1896. Author's collection.

---

[152] *Cornu (1897)

8.2 Some secondary but innovative works

After an unusual abundance, Fizeau's creative vein seems to have largely dried up after 1851. However, as written by Cornu[153], "With less brilliance, [his] new researches, however, were of considerable importance, as the characteristic feature of this great physicist was to have reported, long before we had suspected their value, results and methods whose power or precision we now progressively recognize." The main activity of Fizeau, outside of his experiments on the research of the relative motion of the Earth and æther, was now devoted to the application of interferometry to metrology, a field in which he became a master. These were essentially routine works, perhaps conducted initially to pass the time, but which had important practical applications; in them, we recognize the skill and mastery of Fizeau in interferometry, which made him one of the founders of the brilliant French school of metrology.

He started by measuring, by interferometry, the variation with temperature of the refractive index of glass and various other solids, crystalline or not[154]. Fizeau used as a source of monochromatic light that of sodium vapor, obtained with a Brewster lamp in which sea salt was heated in an alcohol flame. Fizeau knew the properties of sodium vapor, which produces the lines of the famous D doublet, in emission or absorption depending on the circumstances as already noticed by Foucault, then by Kirchhoff and Bunsen[155]. He illuminated a thin plate of the material under study with these lines, and observed the interferences between the light reflected by the upper face and that which had traversed twice through the plate after reflection on the lower side (Fig. 8.2). If the plate had been perfect, he would have seen a uniform color, but the imperfections produced irregular interference fringes of equal thickness. If the material was then heated in a furnace, which Fizeau called the "oven" (Fig. 8.3), the fringes moved, and by simply counting how many fringes passed at a given point he could determine, knowing the wavelength of the light, by how much the optical path in the plate had increased. Of course, the fringes given by the two components of the doublet D were superimposed, which complicated the phenomenon. To check the results, Fizeau also occasionally used a lithium lamp that gave only a single line, but was

---

[153] *Cornu (1897), p. C.33
[154] *Fizeau, H. (1862) *CRAS* 54, p. 1237-9 and *Annales de Chimie et de Physique* 66, p. 429-82.
[155] *Fizeau, H. (1862) *CRAS* 54, p. 493-5.

less bright than the sodium lamp and gave a less pure light, because there was sodium mixed with lithium. Of course, the variation of the optical thickness of the plate with temperature was due to both expansion and the variation of the refractive index. Measuring the expansion coefficient by mechanical means, by himself or by others, he could access the second quantity.

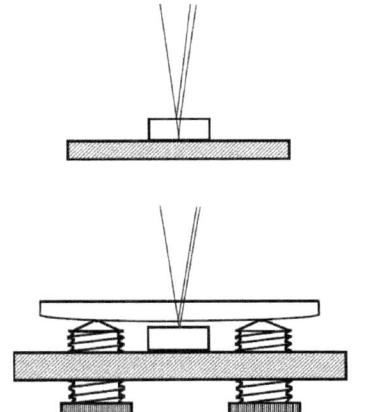

FIGURE 8.2 – Two set-ups used by Fizeau to measure the temperature dependence of the refractive index of a transparent solid (top), and to measure the expansion of a solid (bottom). Top, he observed interferences between the light reflected by the front face and by the rear face of the solid in monochromatic light. Bottom, Newton's rings were observed, formed by interference of the light reflected by the upper face of the solid and that reflected by the lower side of a very slightly convex plate, whose position was adjusted by screws. Author's drawing.

FIGURE 8.3 – The "oven" used by Fizeau for his experiments on the variation of the refractive index with temperature and the expansion of solids. The material under study, and later the "tripod" of Figure 8.4, was placed in the oven, its temperature being measured by two thermometers. The bulb of one of the thermometers was in contact with the material, and the observation of the scale was made by a lens system and a 45° mirror visible on the top. The diameter of the oven is of the order of 10 cm. The oven was fixed by its base on an envelope made by a sheet of lead that surrounded it on all sides, except for a window for the introduction of light and the observation. This envelope was replaced in 1865 by a double copper enclosure. The whole apparatus was heated by an alcohol lamp, and later by two lamps. The thermometers are by Henri Soleil (?-1879). Musée de Suresnes, inv. 997.00.3132.

However, the uncertainty about the expansion coefficients being rather large, Fizeau decided to measure them separately by interference[156]. For this, a plate of the studied material was placed on a steel plate (which he called a "tripod") with three screws that supported a glass plate at a very small distance from the flat surface of the material[157]. The tripod was replaced in 1865 by a new one, of platinum-iridium[158] (Fig. 8.2 and 8.4). By illuminating the assembly with a sodium lamp, Newton's rings were formed by interference between the upper surface of the crystal and the lower surface of the glass plate, which was slightly convex. Fizeau put the whole device into the oven shown in Figure 8.3, and observed the radial displacement of the rings when the temperature varied. This movement being now due to the combined effect of the expansion of the body and steel screws, he had to have the expansion coefficient measured independently. The device could be used to measure the coefficient of expansion of very small samples, for example a diamond, provided they have two parallel faces.

Armed with these instruments, whose improvement can be followed in successive publications, Fizeau measured the coefficient of expansion of many substances, while abandoning the measurements on the changes in the refractive index, which seemed of lesser interest. He eventually used an electrically-excited sodium-vapor lamp (Fig. 8.5) and a mercury vapor lamp (Fig. 8.6). His colleagues, the chemists and mineralogists Henri Sainte-Claire Deville (1818-1881) and Alfred Des Cloiseaux (1817-1897), were enlisted to provide the samples. In 1869, Fizeau presented the expansion coefficients of 40 compounds, pure or composite, crystalline or not, to the Academy of Sciences[159]. One can find manuscript records of experiments in the Archives of the Academy of Sciences dating from 1865 to 1870, which contain the results of measurements of 103 different substances. Fizeau probably found that only some deserved to be published. The program seems to have stopped in 1870, except for measurements of platinum, iridium and their alloys which will be discussed later.

---

[156] *Fizeau, H. (1864) CRAS 58, p. 923-32 and *Annales de Chimie et de Physique 4ᵉ ser. 2, p. 143-85.

[157] The crystals measured by Fizeau were generally "skillfully" polished by Henri Soleil, the son of Jean-Baptiste Soleil (1798-1878) and his successor in 1849. Later, his company twas led by his nephew Leon Laurent and bought in 1893 by Amédée Jobin; it became part of the Jobin Yvon company, currently Horiba Jobin Yvon.

[158] See *Fizeau, H. (1866) CRAS 62, p. 1133-48, and Procès-verbaux de la Commission internationale du mètre, Annales du Conservatoire des Arts et Métiers 10 (1873-7) p. 45, accessibles via http://cnum.cnam.fr .

[159] *Fizeau, H. (1869) CRAS 68, p. 1125-31.

FIGURE 8.4 – Fizeau's "tripod" to measure the coefficient of expansion of a solid by interference. This is his second tripod (1865) in platinum with 10% iridium. The sample, cut so as to have two parallel faces, was placed on the small protuberances arranged in a triangle on the disk, and an optical-quality glass plate was laid on the blunt points of three large screws and adjusted so that it was parallel to the sample surface, and very close to this surface. Locking screws are visible laterally. The diameter of the disk is 3.2 cm. Newton's rings were observed when the sample was illuminated with monochromatic light. The assembly was placed in the oven shown in Figure 8.2 and was therefore heated to a known temperature. It had probably been made, as the first tripod, "with painstaking care" by Émile Brunner (1834-1895) and his brother Leon (1840-1894). Musée de Suresnes, inv. 997.00.3134.

FIGURE 8.5 – A sodium vapor lamp that belonged to Fizeau. The electrodes excite the vapor enclosed in the U-tube; white crystals can be seen in this tube, which must be sodium chloride. Musée de Suresnes, inv. 997.00.2641.

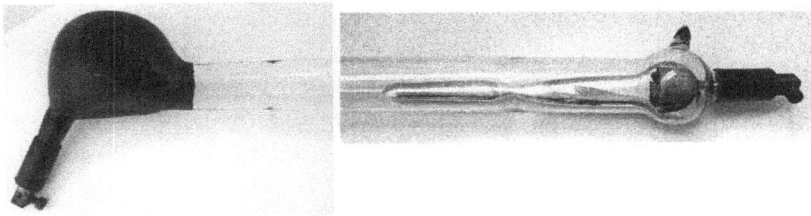

FIGURE 8.6 – A mercury vapor lamp that belonged to Fizeau. Here we see the two ends of the 70-cm long tube, which still contains mercury. Musée de Suresnes, inv. 997.00.2720.

8.3 A pillar of French physics

Since his election to the Academy of Sciences in 1860, Fizeau demonstrated great diligence in this institution and showed much interest in the work of his colleagues. He had an exceptional longevity as a member of the Academy – 36 years, and he was its president for the year 1878. The same year, he was elected as a member of the *Bureau des Longitudes,* an organization created in 1795 that still played an important role in astronomy and navigation, although in 1854 it had lost much of its prestige and its initial prerogatives: before, it had covered the whole of French astronomy. Fizeau was also very assiduous at the meetings of the Bureau. At all these meetings, said Cornu (1897), "He brought, with this impeccable judgment on the conditions needed to obtain correct and accurate results, not only the criticism, often easy, of the proposed projects, but a simple solution of the difficulties raised in the discussion." Indeed, Fizeau had a very sound judgment: when discussing on the 15th of January 1874 at the Council of the Observatory of Paris, to which he belonged, a new astronomical instrument presented by Maurice Loewy (1833-1907), the *coudé equatorial,* he immediately saw the weak point in this telescope, which was designed for the comfort of the observer at the price of two large plane mirrors: he feared that the thermal deformation of these mirrors would degrade the images[160]. The future showed that he was right.

As a member of the Academy of Sciences, Fizeau was particularly interested in metrology, of which he was, along with Michelson, one of the best specialists of the time. In 1872, he was appointed to the *Commission internationale du mètre,* which was responsible for building and distributing the famous standard meters in platinum-iridium (there were about thirty). In 1875, this commission created the International Bureau of Weights and Measures, located at the Pavilion of Breteuil in Sèvres (Hauts-de-Seine), an organization that is still very active. France was obviously very involved in these transactions, which appeared of great political importance, to the extent that the President of the Republic personally attended the melting of the metal for the meters on the 6th of May 1873 and the 1st of May 1874. The construction of these standard meters was supervised by Henri Tresca (1814-1885); Fizeau took an important technical role in this, for example

---

[160] Lequeux, J. (2013) p. 219.

by measuring, with high precision, the coefficient of expansion of the various materials and alloys that were envisaged for the construction of the meters. He also designed methods and devices for the use, measurement and comparison of these bars[161].

Fizeau, who became interested in astronomy in his old age, also played an active role in the Commission for the transit of Venus in 1874. The transits of Venus across the Sun, which are very rare, provide an opportunity to determine the dimensions of the Solar system: indeed, the time of entry of Venus on the solar disk, as well as that of its output, and the place of the disk where these phenomena occur, depend on the observer's position on Earth: this is a parallax effect that depends both on the distance of Venus and of the Sun. Comparing observations made in various places, one can calculate these distances by a method devised in 1716 by Edmond Halley (1656-1742)[162]. The 1874 passage was the first for which photography could be used to observe the event. Fizeau gave precise instructions to obtain these photographs and formed the group of French astronomers that were to be sent on mission. He recommended the use of daguerreotype, which gave better image definition. Numerous expeditions were organized to different parts of the world where the phenomenon was to be visible: 20 English ones, 8 American, 7 Australian, 5 German, etc. As for France, it sent 7 expeditions[163]: in particular, Félix Tisserand (1845-1896) and Jules Janssen (1824-1907) went to Nagasaki in Japan, where Janssen used his "photographic revolver", an ancestor of cinema that allowed him to take thirty successive images of the passage for the first time[164] (Fig. 8.7); Admiral Ernest Mouchez (1821-1892), the future director of the Paris Observatory, (to whom Tisserand succeeded

---

[161] See *(1873 and 1874) *Commission internationale du mètre, réunion des membres français, 1872-1873 et 1873-74, procès verbaux*, Paris, Imprimerie nationale, and the report by Fizeau of the sixth commission of the Commission internationale du mètre, *Annales du Conservatoire des Arts et Métiers* 10 (1873-7) p. 86-8, accessible via http://cnum.cnam.fr .

[162] Halley, E. (1716) A New Method of Determining the Parallax of the Sun, or His Distance from the Earth, *Philosophical Transactions* 29, p. 454-464. In latin; for an English translation, see http://fr.wikipedia.org/wiki/Transit_de_Vénus note 9.

[163] To Nouméa, to Nagasaki, to India, to Saigon, to the Saint-Paul island in the Indian ocean, to Beijing, and to the Campbell island to the south of New-Zealand, where the Mount Fizeau commemorates the scientist.

[164] See https://www.imcce.fr/langues/en/grandpublic/systeme/promenade-en/pages6/608.html. All the texts relative to the passages of Venus are collected in a CD, *Les rendez-vous de Vénus* (2004), distributed by the Institut de Mécanique Céleste et de Calculs des Éphémérides (IMCCE) at the Paris Observatory.

in this function after his death), directed another expedition to the St. Paul Island, a tiny island in the *French Southern and Antarctic Territories*, lying 1,300 km NNE of the Kerguelen islands. The photographic observations, usually on daguerreotypes plates, were successful (Fig. 8.8). Is that because of the instruction of Fizeau? Cornu wrote[165]: "It is thanks to his persevering care that the expedition [which one?[166]] has succeeded in the photographic observation of the phenomenon: and the success would had been even more remarkable if his instructions had been completely followed."

Fig. 5. — Revolver photographique de M. Janssen. — Vue de l'appareil en fonctionnement pendant le passage de Vénus.

FIGURE 8.7 – Janssen observed the transit of Venus across the Sun in Nagasaki with his "photographic revolver". The light was reflected to the camera by a Silbermann heliostat similar to that shown at the beginning of Chapter 2. Bibliothèque de l'Observatoire de Paris.

---

[165] Cornu (1897), p. C.37.
[166] Mouchez writes in his report that he got "489 photographs worth of the micrometric measurements that we will very soon make under the special direction of Mr. Fizeau": *Mouchez (1875) *CRAS* 80, p. 612-8). André, who was in Nouméa, said that he received advice from Fizeau: *André (1875) *CRAS* 80, p. 1281-6. The other observers did not mention Fizeau.

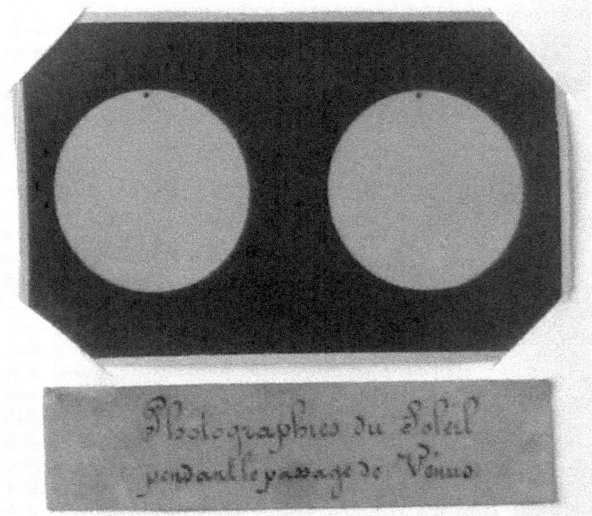

FIGURE 8.8 – Photographs of the transit of Venus across the Sun in 1874 taken by Mouchez at St. Paul Island (paper prints). On the top, one can see the planet silhouetted against the solar disk. Bibliothèque de l'Observatoire de Paris.

8.4 A studious end of life

We have now covered most of the scientific activity of Fizeau. It does not seem very important to list the large number of reports he made to the Academy of Sciences, reports that are accessible through the comprehensive bibliography by Emile Picard[167]. Picard depicts the end of the life of Fizeau well:

"Those, few now amongst us, who knew him here in his late years, can remember the old man with his imposing hair and thick beard, whose first impression was a bit cold. He was only out of his usual reserve when concerned by the interests of Science; although he was enemy of controversies, he became in the discussions an opponent with whom one had to reckon. [...] Very critical for himself, he was wary of some news, and loved to show his respect for those he called the seniors, that is to say for the great mathematicians and the great physicists of the first half of the last century, especially Arago,

---

[167] Picard, E. (1924).

> who had shown him such kindness in his youth, and Fresnel whose work was the constant object of his meditations. Fizeau kept working continuously, and shortly before his death, he was still giving a communication of historical nature on the constancy of the average brightness of the brightest stars. A cruel disease [a cancer of the jaw] removed him after a few weeks from the affection and esteem of the scientific world, on 1896 September 18."

In a world where science was becoming a full-fledged profession, Fizeau was one of the last amateurs: his personal fortune allowed him to conduct his research as he saw fit. He was very famous in his time, as evidenced by the honors and awards he has received and by his elections as a member of numerous foreign academies[168]. If he is less known today, it is because he did not always push his own projects to completion, leaving others to do so (this was also the case of Arago). The most famous follower of Fizeau was Michelson, who was constantly inspired by his ideas. In France, his scientific posterity is found in the great physicists of the optical school that were Cornu, his direct disciple, and later Alfred Perot (1863-1925), Charles Fabry (1867-1945) and Henri Buisson (1873-1944). More recently, we must mention Pierre-Michel Duffieux (1891-1976), the founder of Fourier optics, Pierre Jacquinot (1910-2002) and André Maréchal (1916-2007), to mention only the missing. Thus the optical work of Fizeau assures him an honored place in the history of this science, and his fertile ideas are still sources of inspiration for physicists.

---

[168] Distinctions: Knight of the French *Légion d'honneur*, 1850, then Officer, 1875. Commander of the Order of the Rose of Brazil, 1872. Prizes: *Société d'encouragement pour l'industrie nationale*, 1848 (1,000 francs); Triennial prize of the five Academies, 1856 (30,000 francs); Rumford medal of the Royal Society of London for 1866 (about 1,900 francs). Election to foreign scientific societies: Academy of Berlin, 1863; Pontifical Academy of the Nuovi Lincei, 1866; Royal scientific society of Uppsala, 1870; Royal Society of London, 1875; Royal Academy of sciences of Sweden, 1877; New York Academy of sciences, 1879; Physical society of London, 1880; Academy of sciences of Bologna, 1885; Society of the naturalists of Moscow, 1887; Dutch Society of sciences, 1888; Imperial Society of the friends of science of Moscow, 1888; Royal Academy of Belgium, 1890; Royal Society of Edimburg, 1892; Also Imperial Society of natural sciences of Cherbourg (France), 1865.

Appendix 1. Genealogy of Fizeau and his wife

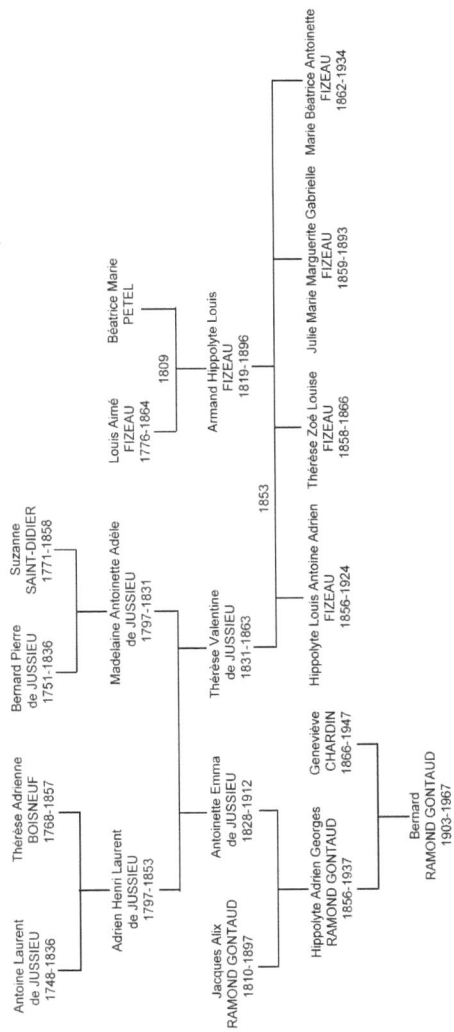

The 7 brothers and sisters of Hippolyte Fizeau are not indicated on the family tree. Those are:

Marie-Rosalie (1811-1854), wife of Auguste Félix Bruzard (1796-1855)

Pierre Joseph Louis (1812-)

Marie Pauline Béatrix (1813-)

Jeanne-Marie (1818-1880), wife in 1839 of Jean Marie Jules Picque (1813-1863)

Marie Gabrielle (1820-1862)

Jules Jean Marie (1822-1880)

Marie François Xavier (1824-1885)

Appendix 2. Chronology

1819, 23 September: Birth of Fizeau in Paris

1835-1842: Studied at *Collège Stanislas* then at the Faculty of medicine

1843-44: Worked on the daguerreotype, encountered Léon Foucault

1845, 2 April: Daguerreotype of the Sun, with Foucault

1845-6: Interferences with a large path difference, with Foucault

1847: Measurement of wavelengths in the infrared, with Foucault

1848, 23 December: Presentation of the Doppler-Fizeau effect at the *Société philomathique*, to which he was elected the following 27 January

1849, February-July: Measurement of the velocity of light with a toothed wheel

May-October: Measurement of the velocity of electricity

1850, 29 April: Note by Arago at the Academy of sciences on the comparison of the velocity of light in air and water

22-25 April: Final quarrel between Fizeau and Foucault

27 May: Foucault deposits a sealed letter at the Academy, in which he writes that Fizeau and himself have tried without success to measure the drag of the aether by air in motion

17 July: Fizeau succeeds in the comparison of the velocity of light in air and in water, but only 7 weeks after Foucault.

1851, 9 April: First tests of a new apparatus to measure the drag of the aether by a flow of air

22 June: Fizeau shows that the diameter of stars can be measured by interferometry

18 July: The definitive apparatus to measure the drag of the aether is ready, but now adapted to a flow of water

29 September: Fizeau describes his successful experiment showing the "partial drag of aether by a flow of water" at the Academy

1852: Interferometric comparison of the refraction index of dry and wet air, with Arago's large apparatus

14 June: Fizeau deposits a sealed letter at the Academy, in which he describes a project to demonstrate the relative motion of the Earth and æther by photometry

July-August: Corresponding experiments, with a negative result

1853, 3 September: Fizeau marries Thérèse Valentine de Jussieu. He stayed more and more often in the Venteuil castle of the Jussieu at Jouarre (Seine et Marne), while keeping an apartment in Paris

1859: Experiments on the "partial drag of æther" by a transparent solid fixed on the Earth, with very dubious results

1860, 2 January: Elected as a member of the Academy of sciences

1862-70: Worked on the variation of the refraction index with temperature and on the expansion of solids.

1863: Death of Fizeau' wife, who had given him four children

1864-7: Becomes an examiner at *École polytechnique*

1866: Death of his daughter Thérèse Zoé, aged 7

Fizeau receives the Rumford medal of the Royal Society of London

1868: Fizeau publishes his 1851 idea to measure the apparent diameter of stars as a simple remark in a report

1872-4: Work for the *Commission internationale du mètre* and for the passage of Venus

1873: Stephan attempts to measure the apparent diameter of many stars, following the indications of Fizeau

1875: Elected as a foreign associate of the Royal Society of London

1878: Elected at the *Bureau des longitudes*

1881-1884: Fizeau resumes his experiments to measure the relative motion of the Earth and æther by photometry, with negative result

1893: Death of Fizeau's second daughter Julie Marie

1896, 18 September: Death of Fizeau at the Venteuil castle

Appendix 3.
Correspondence Fizeau-Foucault

These letters, which relate to the experiment to compare the velocity of light in air and in water, are in the library of the *Musée d'Histoire Naturelle* in Paris, Ms Jus 108, pièces 60, 60a, 61, 61a and 61b. Fizeau's letters are drafts written in pencil and difficult to read, whereas those of Foucault are written in ink and are very easy to read. My friend William Tobin deciphered and transcribed them and has kindly allowed me to publish them.

22 April 1850

My dear Foucault,

The more I think about our conversation of yesterday, the more I feel above reproach. I am deeply convinced that when you consider things as you do, you take the risk of hurting some sacred rights. Remember therefore simply [?] the course of this affair.

Under the inspiration of Mr Arago's lessons we both realized [?] the importance and the beauty of an experiment. After several discussions about it, we both developed ideas on the subject. Towards the month of July [1849] you gave me your ideas. I told you mine. Yours were based on the use of a rotating mirror with a new process and a new optical arrangement. Mine included two methods, one consisting in the use of a toothed disk, the other in the use of a rotating mirror with an optical arrangement that required two telescopes. I presented my memoir to Mr Arago. I made him share our ideas on the subject and I told him that I would only proceed under his formal invitation. It seemed certain to me that this experiment could not be done without him. Many times since then, he has shown the desire that we deal together with this matter. But it was only the question of me. I was afraid that it would not suit him to have a second collaborator, but I thought that you had as much rights as me to take part in this work. You were sick[169]. I tried to save time. I said that to you more than 5 months ago. Then, I had proposed to you to offer our double cooperation to Mr Arago if he would insist more. You accepted but asked me not to rush. This is what I have done, as you can see.

It is at about the same time that we have spoken many times about how we would do the experiment, we discussed as collaborators the entire optical arrangement.

However you had your apparatus built. For the 8 years we have worked together, I have never distrusted you. I thought you intended only to study this new mechanism, to realize a favorite thought and to prepare us to the experiment in question.

Would it be possible now that you are thinking of going alone to the goal, without talking to Mr Arago, and to leave me there as the victim, I might say, of my confidence and my good methods? But Mr Arago has not given on doing the experiment. His right is sacred, but me, what would be my role? Not to mention my rights that you do not dispute, and my methods of which you will not complain; how could I exonerate myself from Mr A.'s allegation that I had let things go without warning him, having sidestepped his proposals? I could only do that by confessing that I had been cheated, and neither you nor I doubt that this would be true. This would be something that seems so contrary to equity, I do not mention friendship, and to your taste that I know for a true scientific glory, pure and honest, that I cannot believe that this project is arrested in your mind. [2 words illegible] and I cannot doubt that you have the desire to do things amicably. Let us go immediately and see Mr Arago.

[H. Fizeau]

---

[169] Indeed, Foucault was depressed and rather inactive during the second semester of 1849.

Paris 23 Avril 1850.

My very dear friend,

Calm down, please; I saw Mr Arago and he listened without anger. I find it absolutely impossible today to talk together; I have too much to take care of the Journal [des Débats] and of you[170].

The more I read your letter, the more I find it extraordinary, incomprehensible; I keep it to reproduce it for you and to give it back to you as soon as you have restored your calm; maybe then you will have some regret.

Meanwhile, you should believe that I am always your friend, much less envious of honors than jealous to live in peace with his conscience and with you.

L. Foucault

Tuesday 24 April 1850 [...but the 24 April 1850 was a Wednesday]

My dear Foucault

I just saw Mr Arago that appeared sensitive to your approach, but I thought you had given up as I have heard nothing more about you. He seems willing to make a statement next Monday at the Academy that will leave the field open. But he seemed very surprised when I asked him his opinion on the facts that I thought were established. He told me they did not seem doubtful and that he and I could not abandon them.

My position in this work would be evident. It is only the question of this experiment on the velocity difference in air and in water, which we have discussed several times as collaborators, and that I am ready to do on my side in a few days. But you understand how such a steeplechase would be unpleasant and painful for both. I will do everything

---

[170] The article of Foucault published in the *Journal des débats* of 24 April is concerned with the measurement of the velocity of electricity by Fizeau and Gounelle, presented on April 15 at the Academy. Foucault is laudatory, but writes: "Young as he still is, Mr. Fizeau has accustomed us to fully accept the facts that he announces; he is his own harshest critic, and when he decided to publish his work, we can only speak highly of it, as it deserves." Should we see a sincere praise there, or is there some irony?

to avoid it, except the sacrifice of what I consider as my right. I know your mind is just enough to be certain that seeing things quietly you will recognize the legitimacy of my claim, and that we will eventually agree as we have always done so far.

Yours faithfully
[H. Fizeau]

Paris 25 April 1850

My dear friend,

I must admit, since you are asking me, that my thoughts are not in your favor; even if I try to argue and to reproduce in all forms your sophistry, I feel that I am preparing for eternal regret. But I am not willing, chained by your tricks, to withdraw the kind of word that I gave you: it would be up to you to untie me. But I feel that on the day of my communication I would find it difficult to conceal the state of my soul. I will try to fix the terms that I would have to use. Still, you may find my pen very restive.

"So, gentlemen, you see that I was absolutely ready; and given the generous declaration of Mr Arago to leave the field open now, I was to solve this wonderful question alone and without scruple, when Mr Fizeau came to suspend my arm and to beg me earnestly not to act without him. The nature of his work had led him to think carefully on this subject and to combine various experiments to remove the difficulties that have not yet been overcome so far. Mr Fizeau had some titles morally in my eyes, and friendship making the rest, we decided that we would approach the first and most pressing application of my method together and that, before doing any measurement, we would share the pleasure and honor of hypothesizing what is the probable difference between the velocities of light in air and in water."

I then fear again, my friend, that you will find all this too true. In this case, let us go to Mr Arago and submit this quarrel to him. But do not bother me this morning[171], I will go and forget the torments you cause me in a lunch with friends.

I affectionately shake your hand.

L. Foucault

---

[171] This suggests that the Fizeau's letter was received on the very morning it had been posted, unless it arrived by carrier.

Bibliography

* indicates the articles or books accessible via http://gallica.bnf.fr

° indicates the articles or books accessible via http://books.google.fr/

⁺ indicates the articles accessible via http://cdsads.u-strasbg.fr

☆ indicates the articles accessible via http://www.jstor.org

Historical books and articles:

*Arago, F. (1854-1857) *Astronomie populaire*, ed. by J.-A. Barral, 4 vol., Paris, Gide & J. Baudry

*Arago, F. (1854-1862) *Œuvres complètes de François Arago*, ed. by J.-A. Barral, 13 vol. Gide, Paris et T. O. Weigel, Leipzig

°Chevalier, C. (1841) *Nouvelles instructions sur l'usage du daguerréotype, description d'un nouveau photographe et d'un appareil très simple destiné à la reproduction au moyen de la galvanoplastie, suivie d'un mémoire sur l'application du brôme.* Paris, chez l'auteur, Palais-Royal et chez Baillière.

*Cornu, A. (1890) Sur la méthode Doppler-Fizeau permettant la détermination par l'analyse spectrale de la vitesse des astres dans la direction du rayon visuel, *Annuaire du Bureau des longitudes pour 1891*, Paris, Gauthier-Villars, p. D.1-D.40.

*Cornu, A. (1897) Notice sur l'œuvre scientifique de H. Fizeau, *Annuaire du Bureau des longitudes pour 1898*, Paris, Gauthier-Villars, p. C.1-C.40.

Fizeau, H. (1859) *Notice sur les travaux de M. H. Fizeau*, Paris, Mallet-Bachelier, accessible via http://www.academie-sciences.fr/activite/archive/dossiers/Fizeau/Fizeau_pdf/Fizeau_notice.pdf

Foucault, L. (1878) *Recueil des travaux scientifiques*, Paris, Gauthier-Villars, rééd. (2001) Paris, Blanchard; accessible via http://jubilotheque.upmc.fr

*Jamin, J.-C. (1885) *Cours de Physique de l'École polytechnique*, t. 3, fasc. 3, Paris, Gauthier-Villars et fils.

°Lerebours & Secrétan (1846) *Traité de photographie, cinquième édition entièrement refondue*, Paris, chez les auteurs, place du Pont-Neuf, et Victor Masson, etc.

*Mascart, E. (1893) *Traité d'optique*, t. 3, Paris, Gauthier-Villars.

°Moigno, F.N.M. (1850) *Répertoire d'optique moderne ou analyse complète des travaux modernes relatifs aux phénomènes de la lumière*, 3e partie, Paris, Franck.

Picard, E. (1924) *Les théories de l'optique et l'œuvre d'Hippolyte Fizeau*, Paris, Gauthier-Villars; contains a very complete list of Fizeau's publications; accessible via http://www.academie-sciences.fr/activite/archive/dossiers/Fizeau/Fizeau_pdf/Fizeau_Picard.pdf

Ranc, A. (1950) Centenaire de la détermination de la vitesse de la lumière par Fizeau, *La Nature,* Août 1950, p. 255-6, accessible via http://cnum.cnam.fr

*Verdet, E. (1872) *Conférences de physique faites à l'École normale*, seconde partie, Paris, Imprimerie Nationale

Recent books and articles:

*Acloque, P. (1984) Hippolyte Fizeau et le mouvement de la Terre: une tentative méconnue, *La vie des sciences, Comptes rendus de l'Académie des sciences* 1, 145-58

Collective (1992) *The Phenomenon of Doppler,* Prague, The Czech Technical University.

*Costabel, P. (1984) L. Foucault et H. Fizeau: exploitation d'une information nouvelle, *La vie des sciences, Comptes rendus de l'Académie des sciences* 1, 235-249

*Costabel, P. (1989) L'entraînement partiel de l'éther selon Fresnel, *La Vie des sciences, Comptes rendus de l'Académie des sciences* 6, p. 327-334

Daniel, M. (1995) *The Beginnings of Photogravure in Nineteenth-Century France*, accessible via http://www.photogravure.com/resources/texts_pdfs/m_daniel_photogravure.pdf.

Eden, Alec (1992) *The Search for Christian Doppler*, Wien, New-York, Springer-Verlag.

Frercks, J. (2000) Creativity and Technology in Experimentation: Fizeaus's Terrestrial Determination of the Speed of Light, *Centaurus* 42, p. 249-87

Frercks, J. (2005) Fizeau's Research Program on Ether Drag. A long Quest for a Publishable Experiment, *Physics in perspective* 7, 3-65

Gough, J.B. (1972) Fizeau, Armand-Hippolyte-Louis, in Gillespie C.C., ed., *Dictionary of Scientific Biography*, vol. 5, p. 18-21.

Labeyrie, A., Lipson, S.G. & Nisenson, P. (2006) *An Introduction to Optical Stellar Interferometry*, Cambridge University Press

Lawson, P. (2006) Optical Interferometry Motivation and History, see http://nexsci.caltech.edu/workshop/2006/talks/Lawson.pdf

Le Gars, S. (2007) *L'émergence de l'astronomie physique en France (1860-1914), acteurs et pratiques*, Thèse, Université de Nantes, accessible via http://tel.archives-ouvertes.fr/docs/00/40/50/48/PDF/These_Stephane_Le_Gars.pdf

Lequeux, J. (2005) *L'univers dévoilé, une histoire de l'astronomie depuis 1910*, Les Ulis, EDP Sciences.

Lequeux, J. (2013) *Le Verrier, Magnificent and Detestable Astronomer*, New York, Springer

Lequeux, J. (2016) *François Arago, a 19th Century French Humanist and Pioneer in Astrophysics*, New York, Springer

Tobin, W. (1993) Toothed Wheels and Rotating Mirrors: Parisian Astronomy and Mid-Nineteenth Century Experimental Measurements of the Speed of Light, *Vistas in Astronomy* 36, p. 253-94.

Tobin, W. (2003) *The Life and Science of Léon Foucault*, Cambridge University Press.

See also the references in https://en.wikipedia.org/wiki/Hippolyte_Fizeau

Index

A

Academy of sciences (French) IV, 1, 3, 4, 6, 10, 20, 22, 37, 43, 47, 56, 66, 69, 82, 91, 98, 116, 120, 122-125
Acloque, Paul 89, 93, 136
Æther 66, 75-96, 117, 118, 129
Ampère, André-Marie (1775-1836) III, 16, 23, 81
Anderson, John A. (1876-1959) 111
André, Charles (1842-1912) 124
Arago, François (1786-1853) III, 1, 3, 4, 7, 10-13, 16, 19, 20, 45-47, 51, 55, 59, 60, 65, 66-72, 75, 76-83, 95, 98, 125, 126, 129, 130, 131-134, 135, 137

B

Belopolsky, Aristarkh (1854-1934) 43
Bessel, Friedrich Wilhelm (1784-1846) 70-71
Biot, Jean-Baptiste (1774-1862) 16
Blanc, Wilfrid 56
Board of longitudes (Bureau des longitudes) 116, 117
Bogaert, Gilles 56
Bolzano, Bernard (1781-1848) 29, 33, 34
Bonnet-Bidaud, Jean-Marc 32
Boussingault, Jean-Baptiste (1802-1887) 2, 3
Breguet, Antoine Louis (1776-1858) 61, 65, 68-71
Breguet, Louis (1804-1883) 71-73
Brewster, David (1781-1868) 17, 118
Brunner, Émile (1834-1895) 121
Brunner, Léon (1840-1894) 121

Buijs-Ballot, Christophorus Hendrik (1817-1890) 33-35, 43
Buisson, Henri (1873-1944) 43
Bunsen, Robert Wilhelm (1811-1899) 11, 39, 118

C

Carnot, Nicolas Léonard Sadi (1796-1832) III
Challis, James (1803-1882) 83
Chevalier, Charles (1808-1895) 4, 6, 7, 10, 135
Coblentz, William W. (1873-1962) 110
Commission internationale du mètre 120-122, 130
Cornu, Alfred (1841-1902) 3, 16, 39, 41, 46, 56, 57, 71, 117, 118, 122, 124, 126, 135
Costabel, Pierre (1912-1989) 79, 80, 82, 136

D

Daguerre, Louis-Mandé (1787-1851) 1, 3, 4, 10
Daguerreotype 2-13, 123-124, 129, 135
Daniel, Malcolm 8, 136
Darrigol, Olivier 94
Davy, Humphry (1778-1829) 10
Des Cloiseaux, Alfred (1817-1897) 120
Deslandres, Henri (1853-1948) 32, 41-42
Diffraction 16, 25
Distinctions, prizes, nominations of Fizeau 126
Doppler, Christian (1803-1853) 28-38, 83, 136

Doppler-Fizeau (effect) 27-44, 95, 129, 135
Drummond, Thomas (1797-1840) 10
Drummond's light (limelight) 9-11, 50, 51, 53
Ducrotay de Blainville, Henri-Marie (1777-1850) 3
Dulong, Pierre-Louis (1785-1838) 3
Dumas, Jean-Baptiste (1800-1884) 2, 3

E

Eden, Alec 28, 136
Einstein, Albert (18879-1955) 42, 75, 96
Electric arc 11, 11, 58
Élie de Beaumont, Léonce (1798-1874) 3
Ether : see Æther
Expansion (measurements by Fizeau) 119-121, 123, 130

F

Fabry, Charles (1867-1945) 43, 126
Faraday, Michael (1791-1867) 61
Faye, Hervé (1814-1902) 38
Fizeau, Marie Gabrielle (1820-1862) 128
Fizeau, Louis-Aimé (1776-1864) 2
Fizeau, Thérèse Zoé (1858-1866) 116, 127, 130
Fourier, Joseph (1768-1830) III, 22, 24, 126
Foucault, Léon (1819-1868) III, IV, 8-13, 15-26, 38, 55, 56, 60, 68-73, 82, 83, 85, 104, 105, 118, 129, 131-137
Fox Talbot, William Henry (1800-1877) 7, 8
Fraunhofer, Joseph von (1787-1826) 18, 25, 33-35, 38
Frercks, Jan 47, 137
Fresnel, Augustin (1788-1827) III, 16, 66, 75, 76, 79-81, 83-85, 95, 96, 126, 136
Froment, Paul Gustave (1815-1865) 45, 51-53, 55, 63, 71

G

Galaxies 32, 44, 105
Galileo Galilei (1564-1642) 47

Galitzin, Boris B. (1862-1916) 43
Galvanometer 63, 90-92
Gough, J.B. 137
Gounelle, Eugène (1821-1864) 48, 61, 63, 133
Gry, Cécile 32
Gurney, Goldsworthy (1793-1875) 10

H

Halley, Edmond (1656-1742) 123
Hamy, Maurice (1861-1936) 109, 110
Hanbury Brown, Robert (1916-2002) 112
Heliostat 12, 15, 50, 53, 73, 88, 124
Herschel, John (1792-1871) 25
Herschel, William (1738-1822) 22-25, 98
Hubble, Edwin (1889-1953) 44
Huggins, William (1824-1910) 39-41, 44
Huygens, Christiaan (1620-1695) 16, 66

I

Induction 61, 62
Infrared 22-25, 33, 110, 114
Interferometry 16-21, 83, 94, 97-114, 118-120, 129, 137
    Intensity interferometry 112, 113
    visibility function 109

J

Jamin, Jules-Célestin (1818-1886) 26, 45, 51, 136
Janssen, Jules (1824-1907) 123, 124
Jupiter 42
    satellites of 109-110
Jussieu, Antoine Laurent de (1748-1836) 116, 117, 127
Jussieu, Bernard Pierre de (1751-1836) 116, 127
Jussieu, Thérèse Adrienne de (1766)1957) 116, 127
Jussieu, Thérèse Valentine de (1831-1863) 116, 127, 130, 135

K

Keeler, James Edward (1857-1900) 42
Kirchhoff, Gustav (1824-1887) 39, 118
Kochenhauer 107
Kreil, Karl (1796-1862) 33

Index

L

Labeyrie, Antoine 112-114, 137
Laborde (abbé) 46
Laënnec, Théophile-René-Hyacinthe (1781-1826) 2
Langevin, Paul (1872-1946) III, 89
Langley, Samuel P. (1834-1906) 24
Laplace, Pierre-Simon (1749-1827) III, 79
Laue, Max von (1879-1960) 75, 95, 96
Lauginie Pierre, 10
Lawson, P. 114, 137
Le Gars, Stéphane 137
Lemaître, Georges (1894-1966) 44
Lequeux, James 81, 110, 114, 122, 137
Lerebours, Noël Paymal (1807-1873) 5, 6, 136
Le Verrier, Urbain (1811-1877) III, 9, 56, 88, 105
Loewy, Maurice (1833-1907) 122
Lorentz, Hendrik Antoon (1853-1928) 94-96

M

Magendie, François (1783-1855) 2, 3
Malus, Étienne Louis (1775-1812) III
Mascart, Éleuthère (1837-1908) 73, 88, 89, 136
Maxwell, John Clerk (1831-1879) 41, 42
Melloni, Macedonio (1798-1854) 23, 24, 26, 90, 92
Metrology 118, 122
Michell, John (1724-1793) 75-77
Michelson, Albert A. (1852-1931) 56-58, 60, 75, 86, 89, 94, 95, 103, 108, 109-114, 122, 126
Mitchel, O.-M. 61
Moigno, abbé (1804-1844) 33-35, 46, 82, 90, 136
Montmartre 47, 50, 51, 53, 54, 85
Morley, Edward W. (1838-1923) 86, 89, 94
Mouchez, Amiral Ernest (1821-1892) 123-125

N

Napoléon III (1808-1873) 9
Nature and theories of light 66-73, 76-81
Newcomb, Simon (1835-1909) 56, 57

Newton, Isaac (1642-1727) 16, 46, 66, 67, 72, 76, 79, 80, 94, 106, 119-121
Niepce, Nicéphore (1765-1833) 3, 4
Niven, C. 41
Nobili, Leopoldo (1784-1935) 90, 92
Nordmeyer, Paul 93

O

Observatory
    European Southern (ESO) 97, 114
    Marseilles 103-105
    Nice 41, 56, 57, 113
    Paris 5, 12, 43, 56, 57, 68, 72, 77, 105, 107-109, 122, 123
Orfila, Mathieu Joseph Bonaventure (1787-1853) 2, 3

P

Pease, Francis G. (1881-1938) 103, 110-112
Pelletan, Philippe-Jean (1747-1829) 2
Perrotin, Henri (1845-1904) 56, 57
Petel, Béatrix Marie (?-?), mère de Fizeau 2, 127
Photography 3, 3-12, 42, 90, 123-125
Photometry 10-12, 117, 130
Picard, Émile (1856-1941) 125, 136
Pinel, Philippe (1745-1826) 2
Planck, Max (1858-1947) 110
Poincaré, Henri (1854-1912) III, 41, 42
Polarization of light 12, 19-21, 87, 88

R

Radial velocity 32, 37-44
Radio astronomy 112
Raman, Chandrashekhara Ventaka (1888-1970) 87
Ramond Gontaud, Bernard (1903-1967) 116, 127
Ranc, A. 136
Rayet, Georges (1839-1906) 105
Refraction 16, 25, 48, 66, 67, 76, 77, 81-83, 130
    measurements by Fizeau 118, 119
Regnault, Victor (1810-1878) 2, 3
Relativity 31, 42, 89, 94-96

Repeating circle 77, 78
Rotating mirror : see **spinning mirror**
Ritter, Johannes Wilhelm (1776-1810) 22

S

Sainte-Claire Deville, Henri (1818-1881) 120
Saturn 42
Schwarzschild, Karl (1873-1916) 13, 109, 111
Secchi, P. Angelo (1818-1878) 13, 39, 40, 44
Secrétan, Marc (1804-1867) 5, 136
Sethi, Nihal Karan (1893-1969) 87
Slipher, Vesto M. (1875-1969) 44
Société philomathique 35-37, 43, 129
Soleil, Jean-Baptiste (1798-1878) 120
Soleil, Henri (?-1879) 119, 120
Spectral line 17-19, 25, 33, 34-42
Spectroscopy 17-19, 23-25, 39, 42-44
Spinning mirror 9, 48, 56-61, 65, 66-71, 131, 137
Star 28, 32-39, 43, 44, 76, 77
    Arcturus 110
    Betelgeuse 97, 110-112
    apparent diameter 97-114
    double 28, 32, 34, 41, 117
    Sirius 32, 33, 35, 39, 40, 106-108, 110, 112
    supernova 32
    Vega 112
Stephan, Édouard (1837-1923) 104-108, 130
Stokes, George Gabriel (1819-1903) 83
Studies of Fizeau, 2, 3, 8
Sun 10-14, 19, 22, 23, 25, 33, 35, 39-44
Suresnes IV, 46, 47, 50, 51, 53, 61, 69, 90, 116

T

Thermomultiplier 90, 92
Thollon, Louis (1829-1887) 41

Tisserand, Félix (1845-1896) 123
Tobin, William IV, 15, 17, 18, 50, 72, 73, 131, 137
Toothed wheel 36, 45-54, 56, 61, 62, 69, 129, 131, 137
Tresca, Henri (1814-1885) 122
Twiss, Richard Q. (1925-2005) 112

U

Ultraviolet 11, 24, 43

V

Vauquelin, Louis-Nicolas (1763-1829) 2
Velocity of electricity 46, 58-63, 129, 133
Velocity of light III, 9, 41, 45-58, 77, 129
    in air and water 65-73, 129, 131-134
Venteuil castle 93, 116, 117, 130
Verdet, Émile (1824-1866) 55, 63, 136
Venus 37
    passages of Venus 123-125, 130

W

Walker, S.-C. 61
Watson, William (1715-1787) 58
Wheatstone, Charles (1802-1875) 48, 59-62, 66, 68
Wilip, Josef P. 43
Wolf, Charles (1927-1918) 56
Wollaston, William Francis (1731-1815) 22

Y

Young, Charles Augustus (1834-1908) 41
Young, Thomas (1773-1829) 16, 22, 66

Z

Zeeman, Pieter (1865-1943) 87

www.ingramcontent.com/pod-product-compliance
Ingram Content Group UK Ltd.
Pitfield, Milton Keynes, MK11 3LW, UK
UKHW021329180426
11947UKWH00017B/1535